James Monroe Ingalls

Exterior Ballistics in the Plane of Fire

James Monroe Ingalls

Exterior Ballistics in the Plane of Fire

ISBN/EAN: 9783337250539

Printed in Europe, USA, Canada, Australia, Japan

Cover: Foto ©berggeist007 / pixelio.de

More available books at **www.hansebooks.com**

Department of Ballistics of the U. S. Artillery School.

EXTERIOR BALLISTICS

IN THE

PLANE OF FIRE.

BY

JAMES M. INGALLS,

CAPTAIN FIRST ARTILLERY, U. S. ARMY,
INSTRUCTOR.

NEW YORK:

D. VAN NOSTRAND, PUBLISHER,

23 MURRAY AND 27 WARREN STREETS.

1886,

HEADQUARTERS UNITED STATES ARTILLERY SCHOOL.

FORT MONROE, VA., February, 1885.

Approved and Authorized as a Text-Book.

Par. 26, Regulations U. S. Artillery School, approved 1882, *viz.:*

"To the end that the school shall keep pace with professional progress, it is made the duty of Instructors and Assistant-Instructors to prepare and arrange, in accordance with the Programme of Instruction, the subject-matter of the courses of study committed to their charge   The same shall be submitted to the Staff, and, after approval by that body, the matter shall become the authorized text-books of the school, be printed at the school, issued, and adhered to as such."

BY ORDER OF LIEUTENANT-COLONEL TIDBALL.

TASKER H. BLISS,
First Lieutenant 1st Artillery, Adjutant.

PREFACE.

THIS work is intended, primarily, as a text-book for the use of the officers under instruction at the U. S. Artillery School, and the arrangement of the matter has been made with reference to the wants of the class-room. The aim has been to present in one volume the various methods for calculating range-tables and solving important problems relating to trajectories, which are in vogue at the present day, developed from the same point of view and with a uniform notation. The convenience of this is manifest.

It is hoped, also, that the practical artillerist will find here all that he may require either for computing range-tables for the guns already in use, or for determining in advance the ballistic efficiency of those which may be proposed in the future.

ERRATA.

Page 54, line 27 :

$$\text{For } \frac{1}{u} \text{ read } \frac{1}{v}.$$

Page 64, line 4 :

For (i) and (φ) read $(i)_n$ and $(\varphi)_n$.

Page 72, line 18 :

For $\sec^4 \varphi$ read $\sec^{\frac{4}{5}} \varphi$.

Page 73, line 22 :   .

$$\text{For } \frac{G\,C}{A} \text{ read } \frac{g\,C}{A}.$$

Page 93, line 11 :

For g read y.

Page 116, equation (78) :

$$\text{For } \frac{C}{\cos^3 \varphi} \text{ read } \frac{C}{2 \cos^3 \varphi}$$

CONTENTS.

CONTENTS.

EXTERIOR BALLISTICS

IN THE PLANE OF FIRE.

INTRODUCTION.

Definition and Object.—Ballistics, from the Greek βαλλω, *I throw*, is, in its most general signification, the science which treats of the motion of heavy bodies projected into space in any direction; but its meaning is usually restricted to the motion of projectiles of regular form fired from cannon or small arms.

The motion of a projectile may be studied under three different aspects, giving rise to as many different branches of the subject, called respectively *Interior Ballistics*, *Exterior Ballistics*, and *Ballistics of Penetration*.

1. **Interior Ballistics.**—Interior Ballistics treats of the motion of a projectile within the bore of the gun while it is acted upon by the highly elastic gases into which the powder is converted by combustion. Its object is to determine by calculation the velocity of translation and rotation which the combustion of a given charge of powder of known constituents and quality is capable of imparting to a projectile, and the effect upon the gun.

2. **Exterior Ballistics.**—Exterior Ballistics considers the circumstances of motion of a projectile from the time it emerges from the gun until it strikes the object aimed at. Its data are the shape, caliber, and weight of the projectile, its initial velocity both of translation and of rotation,

the resistance it meets from the air, and the action of gravity.

3. **Ballistics of Penetration.**—This branch of the subject has reference to the effect of the projectile upon an object; the data being the energy and inclination with which the projectile strikes the object, the nature of the resistance it encounters, etc.

The above is not the order in which the three divisions of the subject are usually presented to the practical artillerist, but the reverse. He desires to penetrate or destroy a given object—say the side of an armored ship. Ballistics of penetration enables him to determine the minimum energy which his projectiles must have on impact, and the proper striking angle, to accomplish the desired result. Exterior Ballistics would then carry the data from the object to be struck to the gun, and determine the necessary initial velocity and angle of elevation. Lastly, Interior Ballistics would ascertain the proper charge and kind of powder to be used to give the projectile the initial velocity demanded.

The following pages treat only of Exterior Ballistics; and this subject will be limited, at present, to motion in the vertical plane passing through the axis of the piece.

CHAPTER I.

Preliminary Considerations.—The molecular theory of gases is not yet sufficiently developed to be made the basis for calculating the resistance which a projectile experiences in passing through the air. We know, however, that if a body moves in a resisting medium, fluid or gaseous, the particles of the fluid must be displaced to allow the body to pass through; and hence momentum will be communicated to them, which must be abstracted from the moving body. From the assumed equality of momenta lost and gained Newton deduced the law of the *square* of the velocity to express the resistance of the air to the motion of a body moving in it.

The following, which is the ordinary demonstration, supposes the particles of air against which the body impinges to be at rest, and takes no account of the reaction of the molecules upon each other, nor of their friction against the surface of the body. The result will therefore be but an approximation, which must be estimated at its true value by means of well-devised and accurately-executed experiments.

Normal Resistance to the Motion of a Body presenting a Plane Surface to the Medium.—Let a moving body present to the particles of a fluid against which it impinges, and which are supposed to be at rest, a plane surface whose area is S, and which is normal to the direction of motion. Let w be the weight of the moving body, v its velocity at any time t, δ the weight of an unit-volume of the fluid, and g the acceleration of gravity. The plane S will describe in an element of time dt a path $v\,dt$, and displace a volume of fluid $S\,v\,dt$; therefore the mass of fluid put in motion during the element of time is $\dfrac{\delta}{g} S v\,dt$.

And as this moves with the velocity v, its momentum is $\frac{\delta}{g} S v^2 dt$; and this has been abstracted from the moving body, whose velocity has thereby been decreased by dv. Therefore

$$-\frac{w}{g} dv = \frac{\delta}{g} S v^2 dt$$

or

$$-\frac{w}{g} \frac{dv}{dt} = \frac{\delta}{g} S v^2$$

The first member of this last equation is the momentum-decrement of the body, due to the pressure of the fluid upon the plane face S, and is therefore a measure of this pressure. Calling this latter P, we have

$$P = -\frac{w}{g} \frac{dv}{dt} = \frac{\delta}{g} S v^2$$

or, per unit of mass,

$$\frac{g}{w} P = -\frac{dv}{dt} = \frac{\delta}{w} S v^2$$

As before stated, several circumstances have been omitted in this investigation which, if taken into account, would probably increase the pressure somewhat, at least for high velocities. We will therefore introduce into the second member of the above equation an undetermined multiplier k ($k > 1$), and we have

$$P = k \frac{\delta}{g} S v^2 \qquad (1)$$

The pressure is, therefore, proportional to the area of the plane surface, to the density of the medium, and to the square of the velocity.

If in equation (1) we make $S = 1$, the second member will then express the normal pressure upon an unit-surface moving with the velocity v; calling this p_0, we have

$$p_0 = k \frac{\delta}{g} v^2$$

and

$$P = p_0 S$$

Oblique Motion.—If the surface S is oblique to the direction of motion, let ε be the angle which the normal to the plane makes with that direction; and resolve the velocity v into its components $v \cos \varepsilon$, perpendicular, and $v \sin \varepsilon$, parallel, to S. This last, neglecting friction, having no retarding effect, we have for the normal pressure upon S the expression

$$P = k \frac{\delta}{g} v^2 S \cos^2 \varepsilon = p_0 S \cos^2 \varepsilon$$

Poncelet (*Mécanique Industrielle*, 403) cites the following empirical formula for calculating the normal pressure, viz. :

$$P = \frac{2 p_0 S}{1 + \sec^2 \varepsilon} \tag{2}$$

derived by Colonel Duchemin from the experiments of Vince, Hutton, and Thibault. As this expression satisfied the whole series of experiments upon which it was based better than any other that was proposed, we will adopt it in what follows.

Pressure on a Surface of Revolution.—Let $A D B$, Fig. 1, be the generating curve of a surface of revolution, which we will suppose moves in a resisting medium in the direction of its axis, $O A$. If $m\, m'\, m'' = dS$ be an element of the surface, inclined to the direction of motion by the angle $N m v = \varepsilon$, it will suffer a pressure in the direction of the normal $N m$, equal, by (2), to

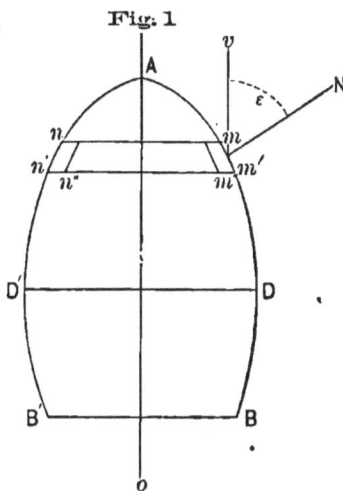

Fig. 1

$$\frac{2 p_0\, dS}{1 + \sec^2 \varepsilon}$$

Resolving this pressure into two components,

$$\frac{2 p_0\, dS \cos \varepsilon}{1 + \sec^2 \varepsilon}, \text{ parallel, and } \frac{2 p_0\, dS \sin \varepsilon}{1 + \sec^2 \varepsilon}, \text{ perpendicular,}$$

to OA, it is plain that this last will be destroyed by an equal and contrary pressure upon the elementary surface $n\,n'\,n''$ situated in the same meridional section as $m\,m'\,m''$, and making the same angle with the direction of motion. It is only necessary, therefore, to consider the first component,

$$\frac{2\,p_0\,d\,S\cos\varepsilon}{1+\sec^2\varepsilon}$$

It is evident that expressions identical with this last are applicable to every element of the zone $m\,m'\,n\,n'$ described by the revolution of $m\,m'$; and we may, therefore, extend this so as to include the entire zone by substituting its area for dS. If we take OA for the axis of X, this area will be expressed by $2\,\pi\,y\,ds$, in which ds is an element of the generating curve; therefore, the pressure upon any elementary zone will be

$$4\,\pi\,p_0\frac{y\,ds\cos\varepsilon}{1+\sec^2\varepsilon}.$$

Substituting $-dy$ for $ds\cos\varepsilon$, and $2+\dfrac{dx^2}{dy^2}$ for $1+\sec^2\varepsilon$, and integrating between the limits $x=l$, and $x=0$, we have

$$P=-2\,\pi\,p_0\int_0^l\frac{y\,dy}{1+\frac{1}{2}\dfrac{dx^2}{dy^2}}$$

As all service projectiles are solids of revolution, this last equation may be used to calculate the relative pressures sustained by projectiles having differently shaped heads, supposing their axes to coincide with the direction of motion at each instant. In applying the formula, y will be eliminated by means of the equation of the generating curve. The superior limit of integration (l) will be the length of the head. R will denote the radius of the projectile.

Application to Conical Heads.—Let $n\,R$ be the length of the conical head, the angle at the point being

$$2\tan^{-1}\left(\frac{1}{n}\right)$$

The equation of the generating line is

$$y = -\frac{x}{n} + R$$

whence

$$\frac{y\,dy}{1 + \frac{1}{2}\frac{dx^2}{dy^2}} = -\frac{2}{n^2(2+n^2)}(nR - x)\,dx$$

and, therefore,

$$P = \frac{4\pi p_0}{n^2(2+n^2)}\int_0^{l=|nR|}(nR - x)\,dx$$

$$= \pi R^2 p_0 \frac{2}{2+n^2}$$

When $n = 0$, the head becomes flat, and the above equation reduces to

$$P = \pi R^2 p_0$$

as it should.

Application to a Prolate Hemi-Spheroidal Head, with Axes in the Ratio of one to two.—The equation of the generating ellipse is

$$4y^2 + x^2 = 4R^2,$$

whence

$$\frac{y\,dy}{1 + \frac{1}{2}\frac{dx^2}{dy^2}} = -\frac{x^3\,dx}{4(8R^2 - x^2)}$$

and, therefore, since $l = 2R$,

$$P = \frac{\pi p_0}{2}\int_0^{2R}\frac{x^3\,dx}{8R^2 - x^2}$$

$$= \pi R^2 p_0(2\log 2 - 1)$$
$$= 0.3863\,\pi R^2 p_0.$$

Application to Ogival Heads.—Let $A\,B\,D$ (Fig. 2) be a section of an ogival head made by a plane passing through the axis of the projectile. Let $A\,O = R$ be the radius of the projectile, and $A\,E = n\,R$ be the radius

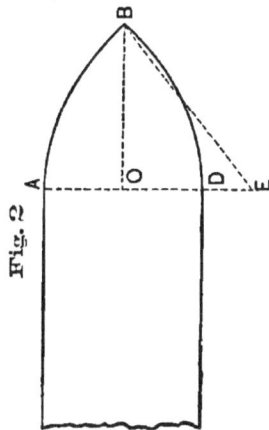

of the generating circle, whose equation is, if we make O the origin and OB the axis of X,

$$y = (n^2 R^2 - x^2)^{1/2} - (n-1) R$$

Making $y = 0$, we find

$$OB = l = R \sqrt{2n-1}$$

Let the angle $AEB = \gamma$; therefore

$$\tan \gamma = \frac{\sqrt{2n-1}}{n-1}$$

which serves to determine the length of the arc of the ogive, AB.

The differential of the equation of the generating circle is

$$dy = - \frac{x \, dx}{(n^2 R^2 - x^2)^{1/2}}$$

whence

$$y \, dy = - x \, dx + \frac{(n-1) R x \, dx}{(n^2 R^2 - x^2)^{1/2}}$$

and

$$1 + \tfrac{1}{2} \frac{dx^2}{dy^2} = \frac{n^2 R^2 + x^2}{2 x^2}$$

therefore

$$P = - 2\pi p_0 \int_0^{R\sqrt{2n-1}} \left\{ \frac{2(n-1) R x^3}{(n^2 R^2 + x^2)(n^2 R^2 - x^2)^{1/2}} - \frac{2 x^3}{n^2 R^2 + x^2} \right\} dx$$

$$= 2\pi R^2 p_0 \left\{ 1 + \frac{n(n-1)}{\sqrt{2}} \log \frac{n + \sqrt{2} + 1}{n - \sqrt{2} + 1} \right.$$

$$\left. - n^2 \log \frac{n^2 + 2n - 1}{n^2} \right\}$$

$$= \pi R^2 p_0 F(n), \text{ (say)} \qquad (3)$$

If a is the angle at the point of the projectile, the expression for dy gives

$$a = 2 \tan^{-1} \left(\frac{\sqrt{2n-1}}{n-1} \right)$$

$$\therefore \gamma = \frac{a}{2}$$

When $n = 1$, $A D B$ becomes a semi-circle and the head a hemisphere.

The following table gives the values of $F(n)$, the lengths of head in calibers, and the angles at the point, for integral values of n from 1 to 6:

n	$F(n)$	LENGTH OF HEAD (l)	ANGLE AT POINT (a)
1	0.6137	0.5000	180° 00′ 00″
2	0.4187	0.8660	120° 00′ 00″
3	0.3176	1.1180	96° 22′ 46″
4	0.2560	1.3229	82° 49′ 09″
5	0.2146	1.5000	73° 44′ 23″
6	0.1848	1.6583	67° 6′ 52″

Resistance of the Air to the Motion of Ogival-headed Projectiles.—The expression

$$P = \pi R^2 p_0 F(n)$$

which, by substituting for p_0 its value, becomes

$$P = k \pi R^2 \frac{\delta}{g} F(n) v^2$$

serves to determine the pressure, as deduced by the above theory, upon an ogival head; and requires that this pressure should be proportional to the density of the air, to the area of the cross-section of the body of the projectile, and to the square of the velocity. The truth of the first two of these deductions may be considered as fully established by experiment, and is admitted by all investigators. The relation between the front pressure and the velocity has not been satisfactorily determined by experiment, and we are therefore unable to verify directly the law of the square deduced above. It seems probable, however, from experiments made to determine the resistance of the air to the motion of projectiles, as well as from theory, that this law is approximately true for all velocities.

If we represent the pressure of the air upon the rear part of the projectile by P', and the resistance by ρ, we shall evidently have

$$\rho = P - P'$$

It is evident that P' will be zero whenever the velocity of the projectile is greater than that of air flowing into a vacuum. In this case, and also when P' is so small relatively to P that it may be neglected, we have approximately

$$\rho = P$$

Application to Ogival Heads struck with Radii of one and a half Calibers.—Experiments have proven that for practicable velocities exceeding about 1300 f. s. the resistance of the air is sensibly proportional to the square of the velocity; and a discussion of the published results of Professor Bashforth's experiments has shown that, within the above limits, the resistance to elongated projectiles having ogival heads struck with radii of one and a half calibers may be approximately expressed by the equation,

$$\rho = \frac{A}{g} d^2 v^2$$

in which d is the diameter of the projectile in inches, g the acceleration of gravity (32.19 ft.), and log $A = 6,1525284 - 10$. Whence

$$\rho = 0.0^544137 \, d^2 v^2$$

Making $\delta = 534.22$ grains, which is the weight of a cubic foot of air adopted by Professor Bashforth, and $F(n) = F(3) = 0.3176$, we find for the corresponding expression for P

$$P = 0.0^541069 \, k \, d^2 v^2$$

A comparison of the second members of these two equations seems to warrant the conclusion that for velocities greater than about 1300 f. s., the rear pressure is either zero or so small relatively to the front pressure that it may be neglected without sensible error. Equating the two members, we find for velocities greater than 1300 f. s.

$$k = 1.0747$$

In the following table the first and second columns give the velocities and corresponding resistances, in pounds, to an elongated projectile one inch in diameter and having an ogival head of one and a half calibers. They were deduced from Bashforth's experiments by Professor A. G. Greenhill, and are taken from his paper published in the Proceedings of the Royal Artillery Institution, No. 2, Vol. XIII. The third column contains the corresponding pressures upon the head of the projectile computed by the formula

$$'P = \frac{\pi \delta F(n)}{576 g} k v^2$$

in which the constants have the values already given. The fourth and fifth columns are sufficiently indicated by their titles.

These results are reproduced graphically in Plate I. A is the curve of resistance (ρ), drawn by taking the velocities for abscissas and the corresponding resistances, in pounds, for ordinates. This curve is similar to that given by Professor Greenhill in his paper above cited. B is the curve of front pressures (P), and is a parabola whose equation is given above. It will be seen that while the velocity *decreases* from 2800 f. s. to 1300 f. s., the two curves closely approximate to each other; the differences $(P - \rho)$ for the same abscissas being relatively small and alternately plus and minus. As the velocity still further *decreases*, the curve of resistance falls rapidly below the parabola B, showing that the resistance now decreases in a higher ratio than the square of the velocity. This continues down to about 800 f. s., when the parabolic form of the curve is again resumed, but still below B. The differences $P - \rho$ from $v = 1300$ f. s. to $v = 100$ f. s. are shown graphically by the curve C, which may represent, approximately, the rear pressures for *decreasing velocities*, and possibly account, in a measure, for the sudden diminution of resistance in the neighborhood of the velocity of sound.

v	ρ	P	$P-\rho$	$\dfrac{P-\rho}{P}$	v	ρ	P	$P-\rho$	$\dfrac{P-\rho}{P}$
2800	35.453	34.603	−0.850		1080	3.999	5.148	+1.149	0.223
2750	33.586	33.378	−0.208		1070	3.756	5.053	1.297	0.256
2700	31.846	32.176	+0.330		1060	3.478	4.959	1.481	0.298
2650	30.241	30.995	+0.754		1050	3.139	4.866	1.727	0.355
2600	28.613	29.836	+1.223		1040	2.823	4.774	1.951	0.409
2550	27.243	28.700	+1.457		1030	2.604	4.684	2.080	0.444
2500	26.406	27.585	+1.379		1020	2.482	4.592	2.114	0.459
2450	25.898	26.493	+0.595		1010	2.404	4.502	2.098	0.466
2400	25.588	25.422	−0.166		1000	2.330	4.414	2.084	0.472
2350	25.242	24.374	−0.868		990	2.261	4.326	2.065	0.477
2300	24.760	23.347	−1.413		980	2.193	4.239	2.046	0.483
2250	23.566	22.344	−1.222		970	2.127	4.153	2.026	0.488
2200	22.158	21.362	−0.796		960	2.061	4.068	2.007	0.493
2150	20.811	20.402	−0.409		950	1.998	3.983	1.985	0.498
2100	19.504	19.464	−0.040		940	1.935	3.900	1.965	0.504
2050	18.229	18.548	+0.319		930	1.874	3.817	1.943	0.509
2000	17.096	17.654	+0.558		920	1.814	3.736	1.922	0.515
1950	16.127	16.783	+0.656		910	1.756	3.655	1.899	0.520
1900	15.364	15.934	+0.570		900	1.699	3.575	1.876	0.525
1850	14.696	15.106	+0.410		850	1.431	3.189	1.758	0.551
1800	14.002	14.300	+0.298		800	1.212	2.825	1.613	0.580
1750	13.318	13.517	+0.199		750	1.043	2.483	1.440	0.580
1700	12.666	12.766	+0.100		700	0.905	2.163	1.258	0.581
1650	12.030	12.016	−0.014		650	0.784	1.865	1.081	0.580
1600	11.416	11.298	−0.018		600	0.674	1.589	0.915	0.576
1550	10.829	10.604	−0.225		550	0.572	1.335	0.763	0.572
1500	10.263	9.930	−0.333		500	0.473	1.103	0.630	0.571
1450	9.622	9.280	−0.342		450	0.381	0.894	0.513	0.574
1400	8.924	8.651	−0.273		400	0.294	0.706	0.412	0.583
1350	8.185	8.044	−0.141		350	0.221	0.541	0.320	0.592
1300	7.413	7.459	+0.046	0.006	300	0.162	0.397	0.235	0.592
1250	6.637	6.896	0.259	0.038	250	0.112	0.276	0.164	0.595
1200	5.884	6.356	0.472	0.070	200	0.072	0.177	0.105	0.591
1150	5.179	5.837	0.658	0.113	150	0.040	0.099	0.059	0.594
1100	4.420	5.340	0.920	0.172	100	0.018	0.044	+0.026	0.591
1090	4.221	5.244	+1.023	0.195					

CHAPTER II.

EXPERIMENTAL RESISTANCE.

Notable Experiments.—Benjamin Robins was the first to execute a systematic and intelligent series of experiments to determine the velocity of projectiles and the effect of the resistance of the air, not only in retarding but in deflecting them from the plane of fire. He was the inventor of the *ballistic pendulum*, an instrument for measuring the *momenta* of projectiles and thence their velocities. He also invented the *Whirling Machine* for determining the resistance of air to bodies of different forms moving with low velocities. His "New Principles of Gunnery," containing the results of his labors, was published in 1742, and immediately attracted the attention of the great Euler, who translated it into French.

The next series of experiments of any value were made toward the close of the last century by Dr. Hutton, of the Royal Military Academy, Woolwich. He improved the apparatus invented by Robins, and used heavier projectiles with higher velocities. His experiments showed that the resistance is approximately proportional to the square of the diameter of the projectile, and that it increases more rapidly than the square of the velocity up to about 1440 f. s., and nearly as the square of the velocity from 1440 f. s. to 1968 f. s.

In 1839 and 1840 experiments were conducted at Metz, on a hitherto unprecedented scale, by a commission appointed by the French Minister of War, consisting of MM. Piobert, Morin, and Didion. They fired spherical projectiles weighing from 11 to 50 pounds, with diameters varying from 4 to 8.7 inches, into a ballistic pendulum, at distances of 15, 40, 65, 90, and 115 metres; by this means velocities were determined at points 25, 50, 75, and 100 metres apart, the velocities varying from 200 to 600 metres per second.

From these experiments General Didion deduced a law of resistance expressed by a binomial, one term of which is proportional to the square, and the other to the cube, of the velocity. This gave good results for short ranges; but with heavy charges and high angles of projection the calculated ranges were much greater than the observed.

Another series of experiments was made at Metz, in the years 1856, 1857, and 1858, by means of the electro-ballistic pendulum invented by Captain Navez, of the Belgian Artillery. This, unlike the ballistic pendulum, affords the means of measuring the velocity of the same projectile at two points of its trajectory. The results of these elaborate experiments may be briefly stated as follows: The resistance for a velocity of 320 m. s. does not differ sensibly from that deduced from the previous experiments at Metz; but the resistances decrease with the velocity below 320 m. s., and increase with the velocity above 320 m. s., more rapidly than resulted from the former experiments. The commission having charge of these experiments, whose president was Colonel Virlet, expressed the resistance of the air by a single term proportional to the cube of the velocity for all velocities.

In 1865 the Rev. Francis Bashforth, M.A., who had then been recently appointed Professor of Applied Mathematics to the advanced class of artillery officers at Woolwich, began a series of experiments for determining the resistance of the air to the motion of both spherical and oblong projectiles, which he continued from time to time until 1880. As the instruments then in use for measuring velocities were incapable of giving the times occupied by a shot in passing over a series of successive equal spaces, he began his labors by inventing and constructing a chronograph to accomplish this object, which was tried late in 1865 in Woolwich Marshes, with ten screens, and with perfect success. It was afterwards removed to Shoeburyness, where most of his subsequent experiments were made. He employed rifled guns of 3, 5, 7, and 9-inch calibers, and elongated shot having ogival heads struck with radii of 1½ calibers; also smooth-bore guns of similar calibers for firing spherical shot. From the data derived from these experiments he constructed and published, from time to time, extensive tables connecting space and velocity, and time and velocity, which for accuracy and general usefulness have never been excelled. The first of these tables was published in 1870, and his Final Report, containing coefficients of resistance for ogival-headed shot, for velocities extending from 2800 f. s. to 100 f. s., was published in 1880. These experiments will be noticed more in detail further on.

General Mayevski conducted some experiments at St. Petersburg, in 1868, with spherical projectiles, and in the following year with ogival-headed projectiles, supplementing these latter with the experiments made by Bashforth in 1867 with 9-inch shot. An account of these experiments, with the results deduced therefrom, is given in his "Traité Balistique Extérieure," Paris, 1872.

General Mayevski has recently (1882) published the results of a discussion of the extensive experiments made at Meppen in 1881 with the Krupp guns and projectiles. These latter, though varying greatly in caliber, were all sensibly of the same type, being mostly 3 calibers in length, with an ogive of 2 calibers radius. General Mayevski's results, together with Colonel Hojel's still more recent discussion of the same data, will be noticed again.

Methods of Determining Resistances.—If a projectile be fired horizontally, the path described in the first one or two tenths of a second may, without sensible error, be considered a horizontal right line; and, therefore, whatever loss of velocity it may sustain in this short time will be due to the resistance of the air, since the only other force acting upon the projectile, *gravity*, may be disregarded, as it acts at right angles to the projectile's motion. For example, an 8-inch oblong shell, having an initial velocity of

1400 f. s., will describe a horizontal path, in the first two-tenths of a second after leaving the gun, of 278 ft., while its vertical descent due to gravity will be less than 8 inches. Moreover, if its velocity should be measured at the distance of 278 ft. from the muzzle of the gun, it would be found to be but 1380 f. s., showing a loss of velocity of 20 f. s., due to the resistance of the air.

The relation between the horizontal space passed over by a projectile and its loss of velocity may be determined as follows :

Let w be the weight of the projectile in pounds, V and V' its velocities, respectively, at the distances a and a' from the muzzle of the gun, in feet per second, and g the acceleration of gravity.   The *vis viva* of the projectile at the distance a from the gun is $\dfrac{w\,V^2}{g}$, and at the distance a', $\dfrac{w\,V'^2}{g}$; consequently the loss of *vis viva* in describing the path $a'-a$, is $\dfrac{w}{g}(V^2-V'^2)$; and this, by the principle of *vis viva*, is equal to twice the work due to the resistance of the air.   If the distance $a'-a$ is not too great, say from 100 to 300 ft., according to the velocity of the projectile, it may be assumed that for this distance the resistance will not vary perceptibly; and if ρ is the *mean resistance* for this short portion of the trajectory, we shall have

$$\frac{w}{g}(V^2-V'^2)=2\,(a'-a)\,\rho$$

whence

$$\rho=\frac{w\,(V^2-V'^2)}{2\,g\,(a'-a)}$$

As the resistance of the air is proportional to its density, which is continually varying, it is necessary, in order to compare a series of observations made at different times, to reduce them all to some mean density taken as a standard. If δ is the density of the air at the time the observations are made, and δ, the adopted standard density to which the observations are to be reduced, the second member of the preceding equation should be multiplied by $\frac{\partial_{,}}{\partial}$, which gives

$$\rho = \frac{w\,(V^2 - V'^2)}{2\,g(a' - a)}\,\frac{\partial_{,}}{\partial}$$

We may take for the value of $\partial_{,}$ the weight of a cubic foot of air at a certain temperature and pressure; ∂ will then be the weight of an equal volume of air at the time of making the experiments, as determined by observations of the thermometer, barometer, and hygrometer.

As ρ is the *mean* resistance for the distance $a' - a$, it may be considered proportional to the mean velocity, $v = \frac{V + V'}{2}$; and substituting this in the above expression, it becomes

$$\rho = \frac{w\,v\,(V - V')}{g\,(a' - a)}\,\frac{\partial_{,}}{\partial} \tag{4}$$

By varying the charge so as to obtain different values for V and V', the resistance corresponding to different velocities may be determined, and thence the *law of resistance* deduced.

In order to compare the results obtained with projectiles of different calibers, the resistance per unit of surface (square foot) is taken; and, to make the results less sensible to variations of velocity, Didion proposed to divide the values of ρ by v^2, and compare the quotients (ρ') instead of ρ. Therefore, making $\rho' = \frac{\rho}{\pi\,R^2\,v^2}$, equation (4) becomes

$$\rho' = \frac{w\,(V - V')}{g\,\pi\,R^2\,v\,(a' - a)}\,\frac{\partial_{,}}{\partial} \tag{5}$$

It will be observed that since ρ is divided by v^2, the values of ρ' will be constant when the resistance varies as the square of the velocity; when this is not the case ρ' will evidently be a function of the velocity; or $\rho' = A'\,f(v)$ (suppose), where the constant A', and the form of the function, $f(v)$, are both to be determined.

3

Two assumptions have been made in deducing the expression for ρ, neither of which is exactly correct: 1st, that the resistance can be considered constant while the projectile is describing the short path $a' - a$; and, 2d, that this assumed constant resistance is that due to the mean velocity, v. The nature of the error thus committed may be exhibited as follows:

The exact expression for ρ is

$$\rho = -\frac{w}{g}\frac{dv}{dt} = -\frac{wv}{g}\frac{dv}{ds}$$

Comparing this with (4), it will be seen that we have made

$$\frac{\{V - V'}{a' - a} = -\frac{dv}{ds} \qquad .$$

which is true only when the path described by the projectile is infinitesimal.

To determine the amount of error committed, we can recalculate the values of ρ' by means of the law of resistance deduced from the experiments; and it will be found that in the most unfavorable cases the two sets of values of ρ' will not differ from each other by any appreciable amount. For example, suppose the law of resistance deduced by this method is that of the square of the velocity; what is the exact expression for ρ' in terms of $V - V'$ and $a' - a$? We have

$$\rho' = \frac{\rho}{\pi R^2 v^2} = -\frac{w}{g\pi R^2}\frac{dv}{v\,ds}$$

and therefore

$$\rho'\,ds = -\frac{w}{g\pi R^2}\frac{dv}{v}.$$

whence, integrating between the limits V and V', to which correspond a and a', we have, since ρ' is constant in this case,

$$\rho' = \frac{w}{g\pi R^2 (a' - a)}\log\frac{V}{V'}$$

To test the two expressions for ρ', take the follow

ing data from Bashforth's "Final Report," page 19, round 486:

$V = 2826$ f. s.; $V' = 2777$ f. s.; $w = 80$ lbs.; $R = 4$ in. $= \frac{1}{3}$ ft.;

$V - V' = 49$; $g = 32.191$; $a' - a = 150$ ft., and $v =$

$$\frac{V + V'}{2} = 2801.5.$$

We find $\dfrac{w}{g \pi R^2 (a' - a)} = 0.047463$; and this is a factor in both expressions for ρ'. Therefore, by the approximate method,

$$\rho' = 0.047463 \, \frac{49}{2801.5} = 0.00083$$

and by the exact method,

$$\rho' = 0.047463 \log \frac{2826}{2777} = 0.00084.$$

For a second example, suppose the law of resistance to be that of the cube of the velocity. In this case ρ' varies as the first power of the velocity, or $\rho' = A' v$. Therefore

$$A' ds = - \frac{w}{g \pi R^2} \frac{dv}{v^3}$$

whence

$$A' = \frac{w}{g \pi R^2} \frac{\dfrac{1}{V'} - \dfrac{1}{V}}{a' - a}$$

and

$$\rho' = A' v = \frac{w}{g \pi R^2 (a' - a)} \cdot \frac{v (V - V')}{V V'}$$

Comparing this with (5), it will be seen that (omitting the factor $\dfrac{\hat{a}}{a}$) the two equations are identical, if we assume $v^2 = V V'$; and this is very nearly correct when, as in the present case, $V - V'$ is very small compared with either V or V'.

As an example of this method of reducing observations, the experiments made at St. Petersburg in 1868 by General

Mayevski, with spherical projectiles, have been selected. In these experiments the velocities were determined by two Boulengé chronographs, and the times measured were in every case within the limits of 0."10 and 0."15.

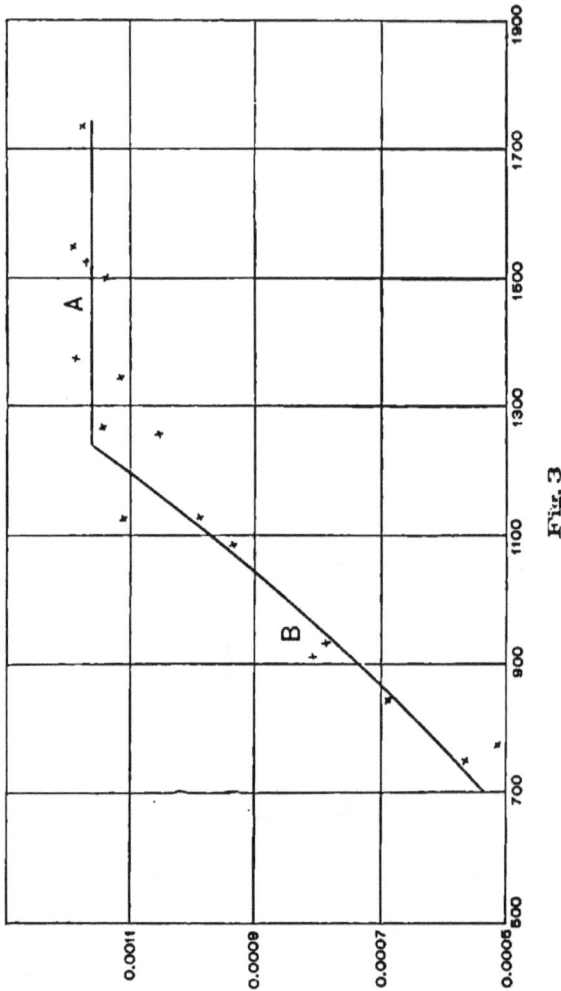

Fig. 3

The experiments were made with 6 and 24-pdr. guns and 120-pdr. mortars, and the velocities ranged from 745 f. s. to 1729 f. s. At least eight shots were fired with the same charge; the value of ρ' was calculated for each shot, and the mean of all the values of ρ' so calculated was taken as corresponding to the mean velocity of all the shots fired with the same charge. The values of $a' - a$ varied from 164 ft. to 492 ft., the least values being taken for the heaviest charges, and the greatest values for the smallest charges. The greatest loss of velocity $(V - V')$ was 131 ft., and the least 33 ft.

The values of ρ' deduced from these experiments are given in the following table. For convenience English units of weight and length are employed; that is, the weights of the projectiles are given in pounds, the velocities in feet per second, and the radii of the projectiles and the values of $a' - a$ in feet.

VALUES OF ρ' FOR SPHERICAL PROJECTILES, DEDUCED FROM THE EXPERIMENTS MADE AT ST. PETERSBURG IN 1868.

Kind of Gun.	Mean Velocity v	Values of ρ'	Kind of Gun.	Mean Velocity v'	Values of ρ'
6-pdr. gun	745 f. s.	0.000561	24-pdr. gun	1247 f. s.	0.001054
24-pdr. gun	768 "	508	o-pdr. gun	1260 "	1145
120-pdr. mort.	860 "	687	120-pdr. mort.	1339 "	1117
6-pdr. gun	912 "	807	6-pdr. gun	1362 "	1189
24-pdr. gun	942 "	782	24-pdr. gun	1499 "	1138
120-pdr. mort.	1083 "	934	120-pdr. mort.	1519 "	1163
24-pdr. gun	1119 "	987	6-pdr. gun	1558 "	1189
6-pdr. gun	1122 "	0.001107	24-pdr. gun	1729 "	0.001178

These results are reproduced graphically in Fig. 3, the velocities being taken for abscissas, and the corresponding values of ρ' for ordinates. It will be seen that the trend of the last seven points is nearly parallel to the axis of abscissas, and may, therefore, be represented approximately by the right line A, whose equation is

$$\rho' = 0.00116$$

in which the second member is the arithmetical mean of the last seven tabulated values of ρ'.

It was found that the remaining points could be best represented by a curve B, of the second degree, of the form $\rho' = p + q v^2$, containing two constants p and q whose values were determined by the method of least squares, each tabular value of ρ' and the corresponding value of v furnishing one "observation equation." It was found that the most probable values of p and q were* $p = 0.012$ and $q = 0.00000034686$; or, reducing to English units of weight and length by multiplying p by $\frac{k}{m^3}$, and q by $\frac{k}{m^5}$, where k is the number of pounds in one kilogramme, and m the number of feet in one metre, we have

$$\rho' = 0.00022832 + 0.0000000061309\, v^2$$

or, in a more convenient form,

$$\rho' = 0.00022832 \left\{ 1 + \left(\frac{v}{610.25}\right)^2 \right\}$$

To find the point of intersection of the right line A with the curve B, equate the values of ρ' given by their respective equations, and solve with reference to v. It will be found that $v = 1233$ f. s., at which velocity we assume that the law of resistance changes.

In strictness there is probably but one *law of resistance*, and this might be, perhaps, expressed by a very complicated function of the velocity, having variable exponents and co-efficients, depending upon the ever-varying density of the air, the cohesion of its particles, etc. ; but, however complicated it may be, we can hardly conceive of its being other than a *continuous* function. But, owing to the difficulties with which the subject is surrounded, both experimental and analytical, it is usual to express the resistance by integral powers of the velocity and constant coefficients, so chosen, as in the above example, as to represent the *mean* resistance over a certain range of velocity determined by experiment.

---

* Mayevski, " Traité de Balistique Extérieure," page 41.

Expression for ρ.—The expression for ρ in terms of ρ' is

$$\rho = \pi R^2 v^2 \rho'$$

which, since ρ' is generally a function of v, may be written

$$\rho = A' \pi R^2 f(v)$$

The resistance per unit of mass, or the retarding force, will therefore be

$$\frac{g}{w}\rho = A' \frac{\pi R^2 g}{w} f(v)$$

or, taking the diameter of the projectile in inches,

$$\frac{g}{w}\rho = A' \frac{\pi g}{576} \frac{d^2}{w} f(v)$$

The first member of this equation expresses the retarding force when the air is at the adopted standard density and the projectile under consideration is similar in every respect to those used in making the experiments which determined ρ'. To generalize the equation for all densities of the atmosphere we must introduce into the second member the factor $\frac{\delta}{\delta_i}$; and we will also assume, at present, that the equation will hold good for different types of projectiles if d^2 be multiplied by a suitable factor (c), depending upon the kind of projectile used. For the standard projectile and for spherical projectiles, $c = 1$; for one offering a greater resistance than the standard, $c > 1$; and if the resistance offered is less, $c < 1$. Making, then,

$$A = A' \frac{\pi g}{576}$$

and

$$C = \frac{\delta_i}{\delta} \frac{w}{c d^2}$$

we have for all kinds of projectiles

$$\frac{g}{w}\rho = -\frac{dv}{dt} = \frac{A}{C} f(v). \qquad (6)$$

C is called the ballistic coefficient, and c the coefficient of reduction.

For the Russian experiments with spherical projectiles the standard density of air to which the experiments were reduced was that of air half saturated with vapor, at a temperature of $15°$ C., and barometer at $0^m.75$. In this condition of air the weight of a cubic metre is $1^k.206$; and, therefore, the weight of a cubic foot ($= \delta_i$) is 0.075283 lbs. $= 526.98$ grs. The value of g taken was $9^m.81 = 32.1856$ feet. Applying the proper numbers, we have the following working expressions for the retarding force for spherical projectiles.

Velocities greater than 1233 f. s.:

$$\frac{g}{w}\rho = \frac{A}{C}v^2; \quad \log A = 6.3088473 - 10$$

Velocities less than 1233 f. s.:

$$\frac{g}{w}\rho = \frac{A}{C}v^2\left(1 + \frac{v^3}{r^3}\right); \quad \log A = 5.6029333 - 10$$

$r = 612.25$ ft.

Oblong Projectiles: General Mayevski's Formulas.—General Mayevski, by a method similar in its general outline to that given above, the details and refinements of which we omit for want of space, has deduced the following expressions for the resistance when the Krupp projectile is employed, viz.: *

$$700^m > v > 419^m, \ \rho = 0.0394\,\pi\,R^2\,\frac{\delta}{\delta_i}\,v^2$$

$$419^m > v > 375^m, \ \rho = 0.0^4 94\,\pi\,R^2\,\frac{\delta}{\delta_i}\,v^3$$

$$375^m > v > 295^m, \ \rho = 0.0^6 67\,\pi\,R^2\,\frac{\delta}{\delta_i}\,v^6$$

$$295^m > v > 240^m, \ \rho = 0.0^4 583\,\pi\,R^2\,\frac{\delta}{\delta_i}\,v^3$$

$$240^m > v > 0^m, \quad \rho = 0.014\,\pi\,R^2\,\frac{\delta}{\delta_i}\,v^2$$

Changing these expressions to the form here adopted

* *Revue d'Artillerie*, April, 1883.

[equation (6)], and reducing to English units of weight and length, they become

$$2300 \text{ ft.} > v > 1370 \text{ ft.}:$$

$$\frac{g}{w}\rho = \frac{A}{C} v^3; \quad \log A = 6.1192437 - 10$$

$$1370 \text{ ft.} > v > 1230 \text{ ft.}:$$

$$\frac{g}{w}\rho = \frac{A}{C} v^3; \quad \log A = 2.9808825 - 10$$

$$1230 \text{ ft.} > v > 970 \text{ ft.}:$$

$$\frac{g}{w}\rho = \frac{A}{C} v^5; \quad \log A = 6.8018436 - 20$$

$$970 \text{ ft.} > v > 790 \text{ ft.}:$$

$$\frac{g}{w}\rho = \frac{A}{C} v^3; \quad \log A = 2.7734232 - 10$$

$$790 \text{ ft.} > v > 0 \text{ ft.}:$$

$$\frac{g}{w}\rho = \frac{A}{C} v^2; \quad \log A = 5.6698755 - 10$$

Colonel Hojel's Deductions from the Krupp Experiments.—Colonel Hojel, of the Dutch Artillery, has also made a study of the Krupp experiments discussed by General Mayevski; and, as it is interesting and instructive to compare the resistance formulas deduced by each of these two experts, both using the same data, we give a brief synopsis of Colonel Hojel's method and results.

He expresses the resistance by the following formula, easily deduced from equation (6):

$$\rho = \frac{R^2}{g} v f(v)$$

in which, from (4),

$$f(v) = \frac{\partial,}{\partial} \frac{w(V - V')}{R^2(a' - a)}$$

It is assumed that the loss of velocity, $V - V'$, is some function of the mean velocity v, which can be expressed approximately, for a limited range of velocity, by a monomial of the form

$$f(v) = A v^n$$

4

in which A and n are constants to be determined. The method of procedure is analogous to that followed in determining ρ', and need not be repeated. Colonel Hojel has considered it necessary to employ fractional exponents, thereby sacrificing simplicity without apparently gaining in accuracy. The results he arrived at are as follows: *

$$700^m > v > 500^m, \quad f(v) = 2.1868 \, v^{0.91}$$
$$500^m > v > 400^m, \quad f(v) = 0.29932 \, v^{1.23}$$
$$400^m > v > 350^m, \quad f(v) = 0.0^4205524 \, v^{3.83}$$
$$350^m > v > 300^m, \quad f(v) = 0.0^721692 \, v^4$$
$$300^m > v > 140^m, \quad f(v) = 0.033814 \, v^{1.6}$$

Substituting these values of $f(v)$ in the equation

$$\frac{g}{w}\rho = \frac{R^2}{w} \, v f(v) = \frac{d^2}{4w} \, v f(v)$$

and reducing the results to English units, that is, taking w in pounds, v in feet, and d in inches, we have as the equivalents of Hojel's expressions, all reductions being made, the following :

$$2300 \text{ ft.} > v > 1640 \text{ ft.} :$$
$$\frac{g}{w}\rho = \frac{A}{C} v^{1.91}; \quad \log A = 6.4211771 - 10$$

$$1640 \text{ ft.} > v > 1310 \text{ ft.} :$$
$$\frac{g}{w}\rho = \frac{A}{C} v^{2.23}; \quad \log A = 5.3923859 - 10$$

$$1310 \text{ ft.} > v > 1150 \text{ ft.} :$$
$$\frac{g}{w}\rho = \frac{A}{C} v^{3.83}; \quad \log A = 0.4035263 - 10$$

$$1150 \text{ ft.} > v > 980 \text{ ft.} :$$
$$\frac{g}{w}\rho = \frac{A}{C} v^4; \quad \log A = 6.8232495 - 20$$

$$980 \text{ ft.} > v > 460 \text{ ft.} :$$
$$\frac{g}{w}\rho = \frac{A}{C} v^{2.6}; \quad \log A = 4.3060287 - 10$$

Comparison of Resistances deduced from the above Formulas.—Making $d = 1$ and $\delta_1 = \delta$, in the above

* *Revue d'Artillerie*, June, 1884.

formulas, gives the resistance in pounds per circular inch at the standard density of the air. Calling this ρ_i, we have

$$\rho_i = \frac{A}{g}v^n$$

The following table gives the values of ρ_i for different velocities according to Mayevski's and Hojel's formulas respectively; and also the same derived from "Table de Krupp," Essen, 1881:

Velocity in feet per sec.	ρ_i According to Mayevski.	ρ_i According to Hojel.	ρ' According to Krupp.	Velocity in feet per sec.	ρ_i According to Mayevski.	ρ_i According to Hojel.	ρ' According to Krupp.
2300	21.629	21.598	21.637	1250	5.807	5.715	5.753
2250	20.699	20.710	20.643	1200	4.899	4.888	4.904
2200	19.789	19.840	19.738	1150	3.960	4.160	3.943
2150	18.900	18.987	18.900	1100	3.171	3.331	3.105
2100	18.031	18.153	17.962	1050	2.513	2.640	2.480
2050	17.183	17.337	17.091	1000	1.969	2.068	2.044
2000	16.355	16.538	16.287	950	1.581	1.749	1.720
1950	15.547	15.757	15.359	900	1.344	1.527	1.486
1900	14.760	14.995	14.611	850	1.132	1.324	1.318
1850	13.993	14.250	13.929	800	0.944	1.138	1.162
1800	13.247	13.523	13.181	750	0.817	0.969	0.983
1750	12.521	12.815	12.500	700	0.712	0.815	0.804
1700	11.816	12.125	11.818	650	0.614	0.677	0.648
1650	11.131	11.453	11.059	600	0.523	0.554	0.514
1600	10.467	10.713	10.400	550	0.439	0.446	0.413
1550	9.823	9.981	9.752	500	0.364	0.351	0.313
1500	9.199	9.277	9.126	450	0.294	0.270	
1450	8.596	8.601	8.490	400	0.232	0.201	
1400	8.014	7.954	7.920				
1350	7.315	7.334	7.238				
1300	6.535	6.641	6.445				

Bashforth's Coefficients.—Professor Bashforth adopted an entirely different method from that just developed to determine the coefficients of resistance, of which we will give an outline, referring for further particulars to his work,* which is well known in this country.

* "Motion of Projectiles," London, 1875 and 1881.

We have $v = \dfrac{ds}{dt}$, whence, differentiating and making s the equicrescent variable,

$$\frac{dv}{dt} = -\frac{ds\,d^2t}{dt^3}$$

and this value of $\dfrac{dv}{dt}$ substituted in (6) gives

$$\frac{g}{w}\rho = -\frac{ds\,d^2t}{dt^3} = \left(\frac{ds}{dt}\right)^3 \frac{d^2t}{ds^3} = v^3\frac{d^2t}{ds^2}$$

From this it follows that if the resistance varied as the cube of the velocity, $\dfrac{d^2t}{ds^2}$ would be constant; and we should have

$$\frac{d^2t}{ds^2} = 2b, \text{ (say)};$$

whence, integrating twice,

$$t = bs^2 + as + c$$

which is the relation between the time and space upon this hypothesis. When the resistance is not proportional to the cube of the velocity, $\dfrac{d^2t}{ds^2}$ in the equation

$$\frac{g}{w}\rho = \frac{d^2t}{ds^2}\,v^3 = 2b\,v^3$$

will be variable, and its value must be so determined by experiment as to satisfy this equation for each value of v. Bashforth's method of deducing these values is briefly as follows :

Ten screens are placed at equal distances (150 feet) apart in the plane of fire, and the exact time of the passage of a projectile through each screen is measured by the Bashforth chronograph. The first, second, third, etc., differences of these observed times are taken, which call d_1, d_2, d_3, etc.

Let s be the distance the projectile has moved from some assumed point to any one of the screens, say the first ;

l the constant distance between the screens; and t_s, t_{s+l}, t_{s+2l}, etc., the observed times of the projectile's passing successive screens. Then from a well-known equation of finite differences we have

$$t_{s+nl}=t_s+nd_1+\frac{n(n-1)}{1\cdot2}d_2+\frac{n(n-1)(n-2)}{1\cdot2\cdot3}d_3+\text{etc.}$$

in which n is an arbitrary variable. Arranging the second member according to the powers of n, we have

$$t_{s+nl}=t_s+n\left(d_1-\frac{1}{2}d_2+\frac{1}{3}d_3-\frac{1}{4}d_4+\text{etc.}\right)$$
$$+n^2\left(\frac{1}{2}d_2-\frac{1}{2}d_3+\frac{11}{24}d_4-\frac{10}{24}d_5+\text{etc.}\right)$$
$$+\quad\text{etc.,}\qquad\qquad\text{etc.,}$$

terms multiplied by the cube and higher powers of n.

Since t is a function of s, we have $t_s=f(s)$ and $t_{s+nl}=f(s+nl)$. Expanding this last by Taylor's formula, we have

$$t_{s+nl}=t_s+\frac{dt_s}{ds}\frac{nl}{1}+\frac{d^2t_s}{ds^2}\frac{n^2l^2}{1\cdot2}+\text{etc.}$$

whence, equating the coefficients of the first and second powers of n in the two expansions of t_{s+nl}, we have

$$l\frac{dt_s}{ds}=d_1-\frac{1}{2}d_2+\frac{1}{3}d_3-\frac{1}{4}d_4+\text{etc.}$$

and

$$l^2\frac{d^2t_s}{ds^2}=d_2-d_3+\frac{11}{12}d_4-\frac{10}{12}d_5+\text{etc.}$$

The first of these equations gives

$$\frac{ds}{dt_s}=v_s=\frac{l}{d_1-\frac{1}{2}d_2+\frac{1}{3}d_3-\frac{1}{4}d_4}$$

and the second

$$\frac{d^2t_s}{ds^2}v_s^3=\frac{g}{w}\rho=\frac{v_s^3}{l^2}\left(d_2-d_3+\frac{11}{12}d_4-\frac{10}{12}d_5+\text{etc.}\right)$$

where v_s is the velocity and $\frac{g}{w}\rho$ the resistance per unit of mass at the distance s from the gun.

As an example take the following experiment made with a 6.92-inch spherical shot, weighing 44.094 lbs., fired from a 7-inch gun.* The times of passing the successive screens were as follows:

Screens.	Passed at, Seconds.	d_1	d_2	d_3
1	2.90068	8431	306	10
2	2.98499	8737	316	10
3	3.07236	9053	326	10
4	3.16289	9379	336	10
5	3.25668	9715	346	10
6	3.35383	10061	356	11
7	3.45444	10417	367	11
8	3.55861	10784	378	
9	3.66645	11162		
10	3.77807			

To find, for example, the velocity at the first screen, we have

$$v_1 = \frac{150}{0.08431 - \frac{1}{2}0.00306 + \frac{1}{3}0.00010} = 1811.4 \text{ f. s.,}$$

and at the seventh screen

$$v_7 = \frac{150}{0.10417 - \frac{1}{2}0.00367 + \frac{1}{3}0.00011} = 1465.3 \text{ f. s.}$$

The retarding forces at the same screens are as follows:

$$\frac{g}{w}\rho_1 = \frac{v_1^2}{(150)^2}(0.00306 - 0.00010) = 0.0000013156 \, v_1^2 = 2b_1 v_1^2$$

and

$$\frac{g}{w}\rho_7 = \frac{v_7^2}{(150)^2}(0.00367 - 0.00011) = 0.0000015822 \, v_7^2 = 2b_7 v_7^2.$$

As these small numbers are inconvenient in practice,

* Bashforth, page 43.

Bashforth substituted for them a coefficient K, defined by the equation

$$K = 2b \frac{\partial_{,}}{\partial} \frac{w}{d^2} (1000)^3.$$

In the experiment selected above the weight of a cubic foot of air was 553.9 grains $= \partial$, while the standard weight adopted was 530.6 grains $= \partial_{,}$. Therefore we have

$$K_{,} = \frac{0.00296}{(150)^2} \times (1000)^3 \times \frac{44.094}{(6.92)^2} \times \frac{530.6}{553.9} = 116.1$$

and

$$K_{,} = \frac{0.00356}{0.00296} K_{,} = 139.6*$$

That is to say, when the velocity of a spherical projectile is 1811.4 f. s., $K = 116.1$; and when its velocity is 1465.3 f. s., $K = 139.6$. By interpolation the values of K, after having been determined for a sufficient number of velocities, are arranged in tabular form with the velocity as argument.

Bashforth determined the values of K by this original and beautiful method for both spherical and ogival-headed projectiles; and for the latter for velocities extending from 2900 f. s. down to 100 f. s. The experiments upon which they were based were made under his own direction at various times between 1865 and 1879, with his chronograph, probably the most complete and accurate instrument for measuring small intervals of time yet invented.

Law of Resistance deduced from Bashforth's K.—It will be seen, by examining Bashforth's table of K for ogival-headed projectiles, that as the velocity decreases from 2800 f. s. down to about 1300 f. s., the values of K gradually increase, then become nearly constant down to about 1130 f. s., then rapidly decrease down to about 1030 f. s., become nearly constant again down to about 800 f. s., and then gradually increase as the velocity decreases, to the

* Bashforth's " Mathematical Treatise," page 97.

limit of the table.   These variations show that the law of
resistance is not the same for all velocities, but that it
changes several times between practical limits.   We may
use Bashforth's K for determining these different laws of
resistance as follows :

We have for the standard density of the air,

$$\frac{g}{w}\rho = 2b\,v^3 = \frac{d^2}{w}\frac{K\,v^3}{(1000)^3} \qquad (7)$$

and

$$\rho' = \frac{576\,\rho}{\pi\,d^2\,v^3}$$

from which we get

$$\rho' = \frac{576\,K\,v}{\pi\,g\,(1000)^3}$$

The values of ρ' have been computed by means of this
formula, for ogival-headed projectiles, from $v = 2900$ f. s. to
$v = 100$ f. s., and their discussion has yielded the following
results:

Velocities greater than 1330 f. s.:

$$\frac{g}{w}\rho = \frac{A}{C}v^2; \quad \log A = 6.1525284 - 10$$

1330 f. s. $> v >$ 1120 f. s. :

$$\frac{g}{w}\rho = \frac{A}{C}v^3; \quad \log A = 3.0364351 - 10$$

1120 f. s. $> v >$ 990 f. s. :

$$\frac{g}{w}\rho = \frac{A}{C}v^6; \quad \log A = 3.8865079 - 20$$

990 f. s. $> v >$ 790 f. s. :

$$\frac{g}{w}\rho = \frac{A}{C}v^3; \quad \log A = 2.8754872 - 10$$

790 f. s. $> v >$ 100 f. s. :

$$\frac{g}{w}\rho = \frac{A}{C}v^2; \quad \log A = 5.7703827 - 10$$

These expressions, derived as they are from Bashforth's

coefficients, give substantially the same resistances for like velocities as those computed directly by means of equation (7). The agreement between the two for high velocities is shown graphically by Plate I., in which A is Bashforth's curve of resistance, while that part of the parabola, B, comprised between the limits $v = 2800$ f. s. and $v = 1330$ f. s., is the curve of resistance deduced from the first of the above expressions. If, however, we compare these expressions with those deduced by Mayevski or Hojel from the Krupp experiments, it will be found that these latter give a less resistance than the former for all velocities.

This is undoubtedly due to the superior centring of the projectiles in the Krupp guns over the English, and to the different shapes of the projectiles used in the two series of experiments, particularly to the difference in the shapes of the heads. The English projectiles, as we have seen, had ogival heads struck with radii of $1\frac{1}{2}$ calibers, while those fired at Meppen had similar heads of 2 calibres, and, therefore, suffered less resistance than the former independently of their greater steadiness.

Comparison of Resistances.—Let ρ and $\rho_{,}$ be the resistances of the air to the motion of two different projectiles of similar forms; w and $w_{,}$ their weights; S and $S_{,}$ the areas of their greatest transverse sections; d and $d_{,}$ their diameters; and D and $D_{,}$ their densities. Then, if we suppose, in the case of oblong projectiles, that their axes coincide with the direction of motion, we shall have from (6) for the same velocity, since S and $S_{,}$ are proportional to the squares of their diameters,

$$\frac{\frac{g}{w}\rho}{\frac{g}{w_{,}}\rho_{,}} = \frac{\frac{S}{w}}{\frac{S_{,}}{w_{,}}} ; \quad \text{and} \quad \frac{\rho}{\rho_{,}} = \frac{S}{S_{,}}$$

that is, for the same velocity the resistances are proportional to the areas of the greatest transverse sections, while the retardations are directly proportional to the areas and in-

5

versely proportional to the weights. For spherical projectiles we have

$$S = \tfrac{1}{4}\pi d^2, \quad S_{\prime} = \tfrac{1}{4}\pi d_{\prime}^2, \quad w = \tfrac{1}{6}\pi d^3 D, \quad \text{and } w_{\prime} = \tfrac{1}{6}\pi d_{\prime}^3 D_{\prime};$$

therefore

$$\frac{\dfrac{g}{w}\prime\prime}{\dfrac{g}{w_{\prime}}\prime_{\prime}} = \frac{d_{\prime} D_{\prime}}{d D}$$

that is, for spherical projectiles the retardations are inversely proportional to the products of the diameters and densities. This shows that for equal velocities the loss of velocity in a unit of time will be less, and, therefore, the range greater, *cæteris paribus*, the greater the diameter and density of the projectile.

As the weight of an oblong projectile is considerably greater than that of a spherical projectile of the same caliber and material, it follows that the retardation of the former for equal velocities is much less than the latter, independently of the ogival form of the head of an oblong projectile which diminishes the resistance still more. Indeed, the retarding effect of the air to the motion of a standard oblong projectile, for velocities exceeding 1330 f. s., is less than for a spherical projectile of the same diameter and weight, and moving with the same velocity, in the ratio of 14208 to 20358. As an example, if d and w are the diameter and weight of a solid spherical cast-iron shot which shall suffer the same retardation as an 8-inch oblong projectile weighing 180 lbs. and moving with the same velocity, we shall have, since we know that a solid shot 14.87 inches in diameter weighs 450 lbs.,

$$d = \frac{(14.87)^3 \times 180 \times 20358}{450 \times 64 \times 14208} = 29.65 \text{ inches}$$

and

$$w = \frac{450 \times (29.65)^3}{(14.87)^3} = 3567 \text{ lbs.}$$

The retarding effect of the air to the motion of projectiles of different calibers but having the same initial velocity and angle of projection, is shown graphically in Fig. 4, which was carefully drawn to scale. *A* is the curve which a projectile would describe *in vacuo*, *B* that actually described by a spherical projectile 14.87 in diameter weighing 450 lbs., and *C* that described by a spherical shot 5.9 inches in diameter

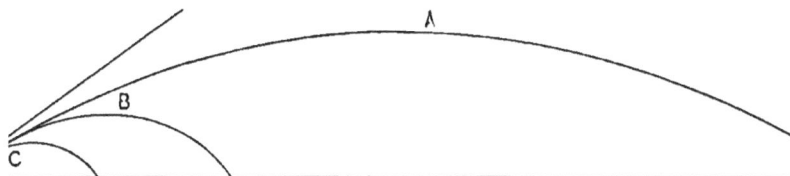

Fig. 4

weighing 26.92 lbs. The initial velocity of each is 1712.6 f. s., and angle of projection 30°.

Example.—Calculate the resistance of the air and the retardation for a 15-inch spherical solid shot moving with a velocity of 1400 f. s. Here $d = 14.87$ in., $w = 450$ lbs., and $A = 20358 \times 10^{-8}$.

Substituting these values in equation (6), we have

$$\rho = \frac{(14.87)^2}{32.16} \times \frac{20358}{10^8} \times (1400)^2 = 2743 \text{ lbs.},$$

and

$$\frac{dv}{dt} = \frac{(14.87)^2}{450} \times \frac{20358}{10^8} \times (1400)^2 = 196.07 \text{ f. s. ;}$$

that is, at the instant the projectile was moving with a velocity of 1400 f. s. it suffered a resistance of 2743 lbs. ; and if this resistance were to remain constant for one second the velocity of the projectile would be diminished by 196.07 ft. As, however, the resistance is not constant, but varies as the square of the velocity, it will require an integration to determine the actual loss of velocity in one second.

We have from (6)

$$\frac{dv}{dt} = -\frac{d^2}{w} A v^2$$

or

$$\frac{dv}{v^2} = - \frac{d^2}{w} A \, dt$$

whence, integrating between the limits V, v, we have

$$v = \frac{V}{1 + A \dfrac{d^2}{w} Vt}$$

Now, making $V = 1400$ and $t = 1$, we find $v = 1228$ f. s.; and the loss of velocity in one second is $1400 - 1228 = 172$ ft.

CHAPTER III.

DIFFERENTIAL EQUATIONS OF TRANSLATION—GENERAL PROPERTIES OF TRAJECTORIES.

Preliminary Considerations.—A projectile fired from a gun with a certain initial velocity is acted upon during its flight only by gravity and the resistance of the air; the former in a vertical direction, and the latter along the tangent to the curve described by the projectile's centre of gravity. It will be assumed, as a first approximation, that the projectile, if spherical, has no motion of rotation; and, in the case of oblong projectiles, that the axis of the projectile lies constantly in the tangent to the trajectory; also that the air through which it moves is quiescent and of uniform density. As none of these conditions are ever fulfilled in practice, the equations deduced will only give what may be called the *normal trajectory*, or the trajectory in the plane of fire, and from which the actual trajectory will deviate more or less It is evident, however, that this deviation from the plane of fire is relatively small; that is, small in comparison with the whole extent of the trajectory, owing to the very great density of the projectile as compared with that of the air.

Notation.—In Figure 5, let O, the point of projection, be taken for the origin of rectangular co-ordinates, of which let the axis of X be horizontal and that of Y vertical. Let $O A$ be the line of projection, and $O B E$ the trajectory described. The following notation will be adopted:

g denotes the acceleration of gravity, which will be taken at 32.16 f. s.;

w the weight of the projectile in pounds;

d its diameter in inches;

φ the angle of projection, $A O E$;

V the velocity of projection, or muzzle velocity ;

U the horizontal velocity of projection $= V \cos \varphi$;

v the velocity of the projectile at any point M of the trajectory ;

ϑ the angle included between the tangent to the curve at any point M and the axis of X, $= T M H$;

ω the angle of fall, $C E O$;

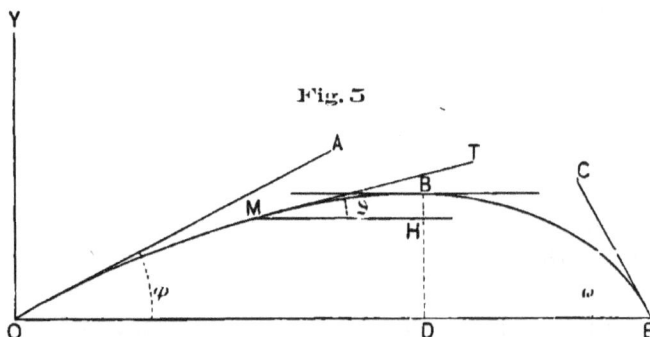

Fig. 5

u the horizontal velocity $= v \cos \vartheta$;

t the time of describing any portion of the trajectory from the origin ;

s the length of any portion of the arc, as $O m$;

X the horizontal range, $O E$;

T the time of flight ;

ρ the resistance of the air, or the resistance a projectile encounters in the direction of its motion, in pounds.

Differential Equations of Translation.—The acceleration* in the direction of motion due to the resistance of the air is $\frac{g}{w}\rho$; and the corresponding acceleration due to gravity is $g \sin \vartheta$; therefore the *total* acceleration in the direction of motion is expressed by the equation,

$$\frac{dv}{dt} = - \frac{g}{w}\rho - g \sin \vartheta \qquad (8)$$

The velocities parallel to X and Y are, respectively,

* The term "acceleration" is here used for retardation. To avoid multiplying terms retardation will be regarded as negative acceleration.

$v \cos \vartheta$ and $v \sin \vartheta$; and the accelerations parallel to the same axes are $\frac{g}{w} \rho \cos \vartheta$ and $g + \frac{g}{w} \rho \sin \vartheta$.

Therefore

$$\frac{d (v \cos \vartheta)}{dt} = -\frac{g}{w} \rho \cos \vartheta \qquad (9)$$

and

$$\frac{d (v \sin \vartheta)}{dt} = -g - \frac{g}{w} \rho \sin \vartheta$$

Performing the differentiations indicated in the above equations, multiplying the first by $\sin \vartheta$ and the second by $\cos \vartheta$, and taking their difference, gives

$$\frac{v \, d \vartheta}{dt} = -g \cos \vartheta \qquad (10)$$

Introducing the horizontal velocity $u = v \cos \vartheta$ in (9) and (10), and substituting for $\frac{g}{w} \rho$ its value from (6), they become, making $f(v) = v^n$,

$$\frac{du}{dt} = -\frac{A}{C} \frac{u^n}{\cos^{n-1} \vartheta} \qquad (11)$$

and

$$\frac{u \, d \vartheta}{dt} = -g \cos^2 \vartheta \qquad (12)$$

whence, eliminating dt,

$$\frac{d \vartheta}{\cos^{n+1} \vartheta} = \frac{g}{A} \frac{C}{u^{n+1}} \frac{du}{u^{n+1}} \qquad (13)$$

Symbolizing the integral of the first member of (13) by $(\vartheta)_n$, that is, making

$$(\vartheta)_n = \int \frac{d \vartheta}{\cos^{n+1} \vartheta}$$

and writing for the sake of symmetry, $\frac{n \, k^n}{g}$ for $\frac{C}{A}$, we shall have

$$(\vartheta)_n = n \, k^n \int \frac{du}{u^{n+1}} = -\frac{k^n}{u^n} + C$$

If (i) is the value of (ϑ) when u is infinite, we have $C = (i)$; and therefore

$$\frac{k^n}{u^n} = (i)_n - (\vartheta)_n \tag{14}$$

whence

$$u = \frac{k}{\left\{ (i)_n - (\vartheta)_n \right\}^{\frac{1}{n}}} \tag{15}$$

and

$$v = \frac{k \sec \vartheta}{\left\{ (i)_n - (\vartheta)_n \right\}^{\frac{1}{n}}} \tag{16}$$

From (11) we have

$$dt = -\frac{C}{A} \cos^{n-1} \vartheta \frac{du}{u^n} \tag{17}$$

and this substituted in the equations

$$dx = u\, dt, \quad dy = u \tan \vartheta\, dt, \quad ds = u \sec \vartheta\, dt,$$

gives

$$dx = -\frac{C}{A} \cos^{n-1} \vartheta \frac{du}{u^{n-1}} \tag{18}$$

$$dy = -\frac{C}{A} \sin \vartheta \cos^{n-2} \vartheta \frac{du}{u^{n-1}} \tag{19}$$

$$ds = -\frac{C}{A} \cos^{n-2} \vartheta \frac{du}{u^{n-1}} \tag{20}$$

From (12) we have

$$dt = -\frac{u}{g} \frac{d\vartheta}{\cos^2 \vartheta} = -\frac{u}{g} d \tan \vartheta \tag{21}$$

whence, as before,

$$dx = -\frac{u^2}{g} d \tan \vartheta \tag{22}$$

$$dy = -\frac{u^2}{g} \tan \vartheta\, d \tan \vartheta \tag{23}$$

$$ds = -\frac{u^2}{g} \sec \vartheta\, d \tan \vartheta \tag{24}$$

Eliminating u from these last four equations by means of (15), they take the following elegant forms:

$$dt = -\frac{k}{g} \frac{d \tan \vartheta}{\left\{ (i)_u - (\vartheta)_u \right\}^{\frac{1}{n}}} \tag{25}$$

$$dx = -\frac{k^2}{g} \frac{d \tan \vartheta}{\left\{ (i)_u - (\vartheta)_u \right\}^{\frac{2}{n}}} \tag{26}$$

$$dy = -\frac{k^2}{g} \frac{\tan \vartheta \, d \tan \vartheta}{\left\{ (i)_u - (\vartheta)_u \right\}^{\frac{2}{n}}} \tag{27}$$

$$ds = -\frac{k^2}{g} \frac{\sec \vartheta \, d \tan \vartheta}{\left\{ (i)_u - (\vartheta)_u \right\}^{\frac{2}{n}}} \tag{28}$$

Remarks.—Subject to the conditions specified in the preliminary considerations, equations (16) to (20) or (25) to (28) contain the whole theory of the motion of translation of a projectile in a medium whose resistance can be expressed by an integral power of the velocity. Equation (16) gives the velocity in terms of the inclination; (18) and (19) or (26) and (27), could they be integrated generally, would give the co-ordinates of any point of the trajectory, while the time would depend upon the integration of (17) or (25). But, unfortunately, the "laws of resistance" which obtain in our atmosphere do not admit of the integration of these equations; we are, therefore, obliged to resort to indirect solutions giving approximations more or less exact. Of these many have been proposed by different investigators; but, with few exceptions, they are either too operose for practical use or not sufficiently approximate.

General Didion, in the fifth section of his "Traité de Balistique," gives a full and interesting *résumé* of the labors of mathematicians upon this difficult problem up to his time (1847), and in the same work gives an original solution of his own of great value. Within the last quarter of a century much has been accomplished to improve and simplify

6

the methods for calculating tables of fire and for the solution of the various problems relating to trajectories; and we will endeavor in the following pages to present such of these methods as are of recognized value, developed after a uniform plan and based upon the preceding differential equations.

General Properties of Trajectories.—Though it is impossible with our present knowledge to deduce the equation of the trajectory described by a projectile, there are certain general properties of such trajectories which may be determined without knowing the law of resistance, if we admit that the resistance increases as some power of the velocity greater than the first, from zero to infinity; whence, making $\frac{\rho}{w} = f(v)$, we shall have $f'(v) > 0$, and $f(\alpha) = \infty$.

Variation of the Velocity—Minimum Velocity. —The acceleration in the *direction of motion* is [equation (8)]

$$\frac{dv}{dt} = -g\,[\,f(v) + \sin \vartheta\,]$$

in which $-g \sin \vartheta$ is the component of gravity in the direction of motion; and, therefore, whether the velocity is increasing or decreasing with the time at any point of the trajectory, depends upon the algebraic sign of the second member; and this, since $f(v) \left(= \frac{\rho}{w} \right)$ is considered positive, depends upon the sign of $\sin \vartheta$. In the ascending branch $\sin \vartheta$ is positive, and, therefore, from the point of projection to the summit the velocity is decreasing. At the summit $\sin \vartheta = 0$, and at this point gravity, which has hitherto conspired with the resistance to diminish the velocity, ceases to act for an instant *in the direction of motion*, and then, as $\sin \vartheta$ changes sign in the descending branch, begins to act in opposition to the resistance; that is, its action tends to increase the velocity. The component of gravity acting perpendicular to the projectile's motion ($g \cos \vartheta$), and which is a maximum at the summit, tends to increase the inclination in the descending branch, and thus to increase (numerically) — sin ϑ, until at a certain point of the descending branch where the inclination is (say) — ϑ' the acceleration of gravity in the direction of motion has increased until it just equals the retardation due to the resistance of the air, which latter has continually decreased with the velocity. Beyond this point, as the component of gravity in the direction of motion still increases with the inclination while the resistance remains constant for an instant, the velocity also increases; and, therefore, at the point where

$$f(v) = \frac{\rho}{w} = -\sin \vartheta'$$

the velocity is a minimum, and $\frac{dv}{dt} = 0$.

Passing the point of minimum velocity, the acceleration of gravity and the retardation due to the resistance of the air both increase; but that there is no maximum velocity, properly speaking, may be shown as follows:

Differentiating the above expression for the acceleration, we have

$$\frac{d^2v}{dt^2} = -g f'(v)\frac{dv}{dt} - g \cos \vartheta \cdot \frac{d\vartheta}{dt}$$

and putting in place of $\frac{d\vartheta}{dt}$ its value from (10), we shall have

$$\frac{d^2v}{dt^2} = -g f'(v)\frac{dv}{dt} + \frac{g^2 \cos^2 \vartheta}{v}$$

and this is necessarily positive whenever $\frac{dv}{dt} = 0$. The velocity, therefore, can only be a minimum; but it tends towards a limiting value, viz., when $\frac{\rho}{w} = 1$, and $\vartheta = -\frac{\pi}{2}$.

Limiting Velocity.—As the limiting velocities of all service spherical projectiles are less than 1233 f. s., we can determine these velocities by means of the expression for the resistance given in Chapter II., from which we get

$$\frac{A}{g}\frac{d^2}{w}\,v^2\left(1+\left(\tfrac{v'}{r}\right)^2\right)=1$$

where $A = 0.000040048$ and $r = 610.25$. Solving with reference to v, we get

$$v = \sqrt{\frac{r^2}{2}\left(\sqrt{1+\frac{4wg}{Ad''r^2}}-1\right)}$$

which gives the limiting velocity.

The following table contains the limiting velocities of spherical projectiles in our service calculated by the above formula :

Solid Shot.	d Inches.	w Lbs.	Final Velocity. Feet.	Shells Unfilled.	d Inches.	w Lbs.	Final Velocity. Feet.
20-inch	19.87	1080	859	15-inch	14.87	330	726
15-inch	14.87	450	783	13-inch	12.87	216	682
13-inch	12.87	283	743	10-inch	9.87	101.75	635
10-inch	9.87	128	684	8-inch	7.88	45	561
12-pdr.	4.52	12.3	526	12-pdr.	4.52	8.34	458

Limit of the Inclination of the Trajectory in the Descending Branch.—We have assumed above that the descending branch of the trajectory ultimately becomes vertical. To prove this, take equation (10), viz.:

$$gdt = -v\frac{d\vartheta}{\cos\vartheta}$$

and integrating from a point of the trajectory where $\vartheta = \varphi$ and $t = 0$, we have

$$gt = \int_\theta^\vartheta v\frac{d\vartheta}{\cos\vartheta}$$

As the velocity v, between the limits $t = 0$ and $t = \infty$, is finite and continuous, and cannot become zero, we have, since v is a function of ϑ,

$$gt = K \int_\theta^\phi \frac{d\vartheta}{\cos \vartheta} = K \log - \frac{\tan\left(\frac{\pi}{4} + \frac{\varphi}{2}\right)}{\tan\left(\frac{\pi}{4} + \frac{\vartheta}{2}\right)}$$

where K is some value of v greater than its least, and less than its greatest value between the limits of integration.

As ϑ is negative in the descending branch, the above equation shows that, when t is infinite, ϑ is equal to $-\frac{\pi}{2}$.

From (24) we have

$$g ds = - v^2 \frac{d\vartheta}{\cos \vartheta}$$

and, therefore, when t is infinite,

$$gs = \int_{\frac{\pi}{2}}^\phi v^2 \frac{d\vartheta}{\cos \vartheta} = K' \log \frac{\tan\left(\frac{\pi}{4} + \frac{\varphi}{2}\right)}{\tan\left(\frac{\pi}{4} - \frac{\pi}{4}\right)} = K' \log \frac{\tan\left(\frac{\pi}{4} + \frac{\varphi}{2}\right)}{\tan 0}$$

where K' is some value of v^2 greater than its least, and less than its greatest value between the limits of inte-

gration; and, as $\log \dfrac{\tan\left(\frac{\pi}{4} + \frac{\varphi}{2}\right)}{\tan 0}$ is infinite, so is the arc which corresponds to $t = \infty$.

Asymptote to the Descending Branch.—As the tangent to the descending branch at infinity is vertical, if it can be shown that it cuts the axis of X at a finite distance, it is an asymptote. To determine this, take equation (22) which gives

$$gx = \int_{-\frac{\pi}{2}}^\phi v^3 \, d\vartheta = K'' \left(\varphi + \frac{\pi}{2}\right)$$

where K'' is a finite quantity, since v^3 is finite between the limits of integration. Therefore the descending branch has a vertical asymptote.

Radius of Curvature.—Designate the radius of curvature by γ. We have by the differential calculus $\gamma = -\dfrac{ds}{d\vartheta}$ (since the trajectory is concave toward the axis of X); we also have $ds = vdt$; consequently $\gamma = -\dfrac{vdt}{d\vartheta}$, and therefore from (12)

$$\gamma = \frac{v^2}{g}\sec\vartheta$$

The radius of curvature is therefore independent of the resistance of the air, and at any point of the trajectory depends only upon the velocity and the inclination, and, therefore, has the same value for the corresponding points of a parabola described by a projectile *in vacuo*. The above expression shows that the radius of curvature decreases from the point of projection to the summit of the trajectory, since v and $\sec\vartheta$ both decrease between those limits. Beyond the summit v still decreases, but as $\sec\vartheta$ increases we cannot determine by simple inspection where γ ceases to decrease and becomes a minimum. Differentiating the expression for γ, we have

$$\frac{d\gamma}{d\vartheta} = \frac{2v\sec\vartheta}{g}\frac{dv}{d\vartheta} + \frac{v^2}{g}\tan\vartheta\sec\vartheta$$

From (13) and (6) we have

$$\frac{d(v\cos\vartheta)}{d\vartheta} = \frac{\rho}{w}v$$

whence, differentiating and reducing,

$$\frac{dv}{d\vartheta} = \frac{v}{\cos\vartheta}\left(\frac{\rho}{w} + \sin\vartheta\right)$$

Substituting this in the expression for $\dfrac{d\gamma}{d\vartheta}$ gives

$$\frac{d\gamma}{d\vartheta} = \frac{v^2}{g}\sec^2\vartheta\left(\frac{2\rho}{w} + 3\sin\vartheta\right)$$

This equation shows that beyond the summit $\dfrac{d\gamma}{d\vartheta}$ is positive up to the point where $\frac{2\rho}{w} + 3 \sin \vartheta = 0$, and then changes its sign. At this point, therefore, the radius of curvature becomes a minimum and afterwards increases to infinity.

At the point of maximum curvature we have, in consequence of the condition $\frac{2\rho}{w} + 3 \sin \vartheta = 0$,

$$\frac{dv}{d\vartheta} = -\frac{1}{2} v \tan \vartheta$$

and therefore, since ϑ is negative in the descending branch, $\frac{dv}{d\vartheta}$ is positive at that point, and v is decreasing with ϑ; in other words, the velocity has not yet become a minimum. Therefore the point of maximum curvature is nearer the summit of the trajectory than the point of minimum velocity.

CHAPTER IV.

RECTILINEAR MOTION.

Relation between Time, Space, and Velocity.—
For many practical purposes, and especially with the heavy, elongated projectiles fired from modern guns, useful results may be obtained by considering the path of the projectile a horizontal right line, and therefore unaffected by gravity. Upon this supposition ϑ becomes zero, and equations (17), (18), and (20) become

$$dt = - \frac{C}{A} \frac{dv}{v^n}$$

and

$$dx = ds = - \frac{C}{A} \frac{dv}{v^{n-1}}$$

whence integrating, and making t and s zero when $v = V$, we have

$$t = C \left\{ \frac{1}{(n-1) A v^{n-1}} - \frac{1}{(n-1) A V^{n-1}} \right\}$$

and

$$s = C \left\{ \frac{1}{(n-2) A v^{n-2}} - \frac{1}{(n-2) A V^{n-2}} \right\}$$

Writing, for convenience,

$$T(v) \text{ for } \frac{1}{(n-1) A v^{n-1}}, \text{ and } S(v) \text{ for } \frac{1}{(n-2) A v^{n-2}}$$

these equations become

$$t = C \{ T(v) - T(v) \} \tag{29}$$

and

$$s = C \{ S(v) - S(v) \} \tag{30}$$

When $n = 2$, the above expression for s becomes indeterminate. In this case we have

$$ds = -\frac{C}{A}\frac{dv}{v}$$

whence

$$s = \frac{C}{A}\left\{ \log V - \log v \right\}$$

and therefore, when $n = 2$,

$$S(v) = -\frac{\log v}{A}$$

Equations (29) and (30) (or their equivalents) were first given by Bashforth in his "Mathematical Treatise," London, 1873. He also gave in the same work tables of $S(v)$ and $T(v)$ for both spherical and elongated shot; the former extending from $v = 1900$ f. s. to $v = 500$ f. s., and the latter from $v = 1700$ f. s. to $v = 540$ f. s. In a "Supplement" to his work above cited, published in 1881, he extended the tables for elongated projectiles to include velocities from 2900 f. s. to 100 f. s.

Projectiles differing from the Standard.—It will be seen that the value of the functions $T(v)$ and $S(v)$ depend upon those of v and A, the former of which is independent of the nature of the shot, while the latter depends partly upon the form of the *standard* projectile, which in this country and England has an ogival head struck with a radius of $1\frac{1}{2}$ calibers, and a body $2\frac{1}{2}$ calibers long. The factor $C\left(\text{or }\frac{\partial_{\prime}}{\partial}\frac{w}{cd^2}\right)$ depends upon the weight and diameter of the projectile, the density of the air, and the coefficient c; which latter varies with the type of projectile used. The factor A varies, therefore, with c; but by the manner in which A and c enter the expressions for t and s, it will be seen that the results will be the same if we make A constant, and give to c a suitable value determined by experiment for each kind of projectile. By this means the tables of the functions $T(v)$ and $S(v)$, computed upon the supposition that $c = 1$, can be used for all types of projectiles. We will now show how these tables may be computed for oblong projectiles, making use of the expressions for the re-

7

sistance derived from Bashforth's experiments given in Chapter I.

Oblong Projectiles, Velocities greater than 1330 f. s.—For velocities greater than 1330 f. s. we have $n = 2$ and $\log A = 6.1525284 - 10$; therefore

$$T(v) = \frac{1}{Av} \text{ and } T(V) = \frac{1}{AV}$$

or, since the value of t depends upon the difference of $T(v)$ and $T(V)$, we may, if convenient, introduce an arbitrary constant into the expression for $T(v)$. Therefore we may take

$$T(v) = \frac{1}{A}\left(\frac{1}{v} + Q_1\right)$$

and, similarly,

$$S(v) = \frac{1}{A}\left(-\log v + \log Q'_1\right) = \frac{1}{A}\log\frac{Q'_1}{v}$$

To avoid large numbers and to give uniformity to the tables we will determine the constants Q_1 and Q'_1 so that the functions shall both reduce to zero for the same value of v; and it will be convenient to begin the table with the highest value of v likely to occur in practice, which we will assume (following Bashforth) to be 2800 f. s.

We therefore have

$$\frac{1}{A}\left(\frac{1}{2800} + Q_1\right) = 0 \qquad Q_1 = -\frac{1}{2800}$$

$$\frac{1}{A}\log\frac{Q'_1}{2800} = 0 \qquad Q'_1 = 2800$$

Substituting the above values of A, Q_1, and Q'_1 in the expressions for $T(v)$ and $S(v)$, and reducing, we have for velocities between 2800 f. s. and 1330 f. s.

$$T(v) = [3.8474716]\frac{1}{u} - .2.5137$$

and

$$S(v) = 55866.12 - [4.2096873]\log v.$$

The numbers in brackets are the logarithms of the numerical coefficients of the quantities to which they are prefixed; and the factor log v is the common logarithm of v, the modulus being included in the coefficient.

Velocities between 1330 f. s. and 1120 f. s.—For velocities between 1330 f. s. and 1120 f. s. we have $n = 3$ and $\log A = 3.0364351 - 10$; therefore, as before,

$$T(v) = \frac{1}{2A}\left(\frac{1}{v^2} + Q_2\right)$$

$$S(v) = \frac{1}{A}\left(\frac{1}{v} + Q'_2\right)$$

Arbitrary Constants.—To deduce suitable values for the arbitrary constants Q_2 and Q'_2, we must recollect that the function representing the resistance of the air changes its form abruptly when the velocity is 1330 f. s.; and to prevent a correspondingly abrupt change in our table at the same point—that is, to make the numbers in the table a continuous series—we must give to Q_2 and Q'_2 such values as shall make the second set of functions equal in value to the first when $v = 1330$. They will, therefore, be determined by the following relations:

$$\frac{1}{2A}\left(\frac{1}{(1330)^2} + Q_2\right) = \frac{1}{A}\left(\frac{1}{1330} - \frac{1}{2800}\right)$$

and

$$\frac{1}{A}\left(\frac{1}{1330} + Q'_2\right) = \frac{1}{A}\ \log\frac{2800}{1330}$$

in which the A in the first member must not be confounded with that in the second. Making the necessary reductions, we have

$$T(v) = [6.6625349]\frac{1}{v^2} + 0.1791$$

and

$$S(v) = [6.9635649]\frac{1}{v} - 1674.1$$

Velocities between 1120 f. s. and 990 f. s.—For velocities between 1120 f. s. and 990 f. s. we have $n = 6$ and $\log A = 3.8865079 - 20$; therefore

$$T(v) = \frac{1}{5A}\left(\frac{1}{v^5} + Q_3\right)$$

and

$$S(v) = \frac{1}{4A}\left(\frac{1}{v^4} + Q'_2\right)$$

The constants must be determined as before, by equating the above expressions to the corresponding ones in the case immediately preceding, making $v = 1120$. The results are, all reductions being made,

$$T(v) = [15.4145221]\frac{1}{v^3} + 2.3705$$

and

$$S(v) = [15.5114321]\frac{1}{v^4} + 4472.7$$

Velocities between 990 f. s. and 790 f. s.—For velocities between 990 f. s. and 790 f. s. we have $n = 3$ and $\log A = 2.8754872 - 10$; whence

$$T(v) = \frac{1}{2A}\left(\frac{1}{v^3} + Q_1\right)$$

and

$$S(v) = \frac{1}{A}\left(\frac{1}{v} + Q'_1\right)$$

Proceeding as before, we have

$$T(v) = [6.8234828]\frac{1}{v^3} - 1.6937$$

and

$$S(v) = [7.1245128]\frac{1}{v} - 5602.3$$

Velocities less than 790 f. s.—For velocities less than 790 f. s. we have $n = 2$ and $\log A = 5.7703827 - 10$; therefore

$$T(v) = \frac{1}{A}\left(\frac{1}{v} + Q_5\right)$$

and

$$S(v) = \frac{1}{A}\log\frac{Q'_5}{v}$$

whence, as before,

$$T(v) = [4.2296173]\frac{1}{v} - 12.4999$$

and

$$S(v) = 124466.4 - [4.5918330]\log v.$$

Ballistic Tables.—Table I. gives the values of the time and space functions for oblong projectiles, computed by the above formulas, and extends from $v = 2800$ f. s. to $v = 400$ f. s. The first differences are given in adjacent columns; and as the second differences rarely exceed eight units of the last order, it will hardly ever be necessary to consider them in using this table.

Table II. gives the values of these functions for spherical projectiles, and is based upon the Russian experiments discussed in Chapter II.

EXAMPLES OF THE USE OF TABLES I. AND II.

Example 1.—The velocity of an 8-inch service projectile weighing 180 lbs. was found by the Boulengé chronograph to be 1398 f. s. at 300 ft. from the gun. What was the muzzle velocity ?

Here $C = \dfrac{180}{64}$, $v = 1398$, and $s = 300$, to find V. From (30) we have

$$S(V) = S(v) - \frac{s}{C}$$

and from Table I.

$$S(1398) = 4903.8 - \frac{3 \times 25.2}{5} = 4888.7$$

also

$$\frac{s}{C} = 300 \times \frac{64}{180} = 106.7$$

whence

$$S(V) = 4782.0$$

$$\therefore V = 1415 + \frac{5 \times 21.6}{24.8} = 1419.4 \text{ f. s.}$$

Example 2.—Determine the remaining velocity and the time of flight of the 12-inch service projectile, weighing 800 lbs., at 1000 yds. from the gun, the muzzle velocity being 1886 f. s.

1. V and s are given, to find v; where $d = 12$, $w = 800$

$V = 1886$, $s = 3000$, and $C = \dfrac{800}{144}$

We have

$$S(v) = S(1886) + \frac{3000 \times 144}{800}$$

From Table I.,

$$S(1886) = 2803.7 - 0.6 \times 37.4 = 2781.3$$

$$\frac{3000 \times 144}{800} = 540.0$$

$$S(v) = 3321\ 3$$

$$\therefore v = 1740 + \frac{10 \times 27.0}{40.3} = 1746.7 \text{ f. s.}$$

2. V and v are given, to find t; from Table I.,

$$T(v) = 1.516$$
$$T(V) = 1.217$$
$$T(v) - T(V) = 0.299$$
$$\therefore t = 0.299 \times \frac{800}{144} = 1''.66$$

Example 3.—Suppose we wish to determine the value of the coefficient of reduction, c, for a particular projectile whose form differs from the standard projectile. From (30) we have

$$C = \frac{w}{c\,d^2} = \frac{s}{S(v) - S(V)}$$

whence

$$c = \frac{w}{d^2} \frac{S(v) - S(V)}{s}$$

It is, therefore, only necessary to measure the velocity of the projectile at two points of its trajectory as nearly in the same horizontal line as practicable, and at a known distance apart, and substitute the values thus obtained in the above formula. For example, the 40-centimetre (71-ton) Krupp gun fires a projectile weighing 1715 lbs. with a muzzle velocity of 1703 f. s. By experiment it is found that the velocity at 1800 ft. from the gun is 1646 f s. What is the value of c for this projectile?

Here $V = 1703$, $v = 1646$, $s = 1800$, $w = 1715$, and $d = 15.748$.

From Table I.,

$$S(v) = 3742.2$$
$$S(V) = \underline{3499.7}$$
$$\log 242.5 = 2.3846580$$
$$\log \frac{w}{d^3} = 0.8397959$$
$$c \log s = 6.7447275$$
$$\log c = 9.9691814 \quad c = 0.9315$$
$$\therefore \log C = 0.8706145$$

Extended Ranges.—For the heaviest elongated projectiles, fired with high initial velocities, the remaining velocities and times of flight may be determined by this method with sufficient accuracy for quite extended ranges; that is to say, for ranges due to an angle of projection of 10° or 12°, or, in other words, when the least value of cos ϑ for the entire trajectory does not depart very much from unity, its assumed value.

Example 4.—Compute the remaining velocities, with the data of the last example, at 1800 ft., 3600 ft., 5400 ft., . . . up to 18000 ft. from the gun.

The work may be arranged as follows:

$$S(v) = 3499.7, \quad \log C = 0.8706145.$$

s	$\dfrac{s}{C}$	$S(v)$	v	v Computed by Krupp's Formula.
1800 ft.	242.47	3742.2	1645 f. s.	1646 f. s.
3600 "	484.9	3984.6	1589 "	1590 "
5400 "	727.4	4227.1	1536 "	1536 "
7200 "	969.9	4469.6	1484 "	1484 "
9000 "	1212.3	4712.0	1434 "	1434 "
10800 "	1454.8	4954.5	1385 "	1385 "
12600 "	1697.3	5197.0	1338 "	1338 "
14400 "	1939.8	5439.5	1293 "	1293 "
16200 "	2182.2	5681.9	1250 "	1251 "
18000 "	2424.7	5924.4	1211 "	1211 "

The numbers in the second column are simple multiples of the first number in that column; those in the third column are found by adding $S(V) = 3499.7$ to the numbers on the same lines in the second column, and the velocities in the fourth column are taken from Table I. with the argument $S(v)$.

The velocities in the last column were computed by Krupp's formula. They are copied, as also the data of the problem, from "Professional Papers No. 25, Corps of Engineers, U. S. A.," page 41.

In this example the angles of projection and fall for a range of 18000 feet are, respectively, 7° 18′ and 9° 20′; while an 8-inch shell weighing 180 lbs. would require for the same range, with the same initial velocity, an angle of projection of 11° 5′, and the angle of fall would be 19° 40′.

In this latter case the velocity computed by the above method would not be a very close approximation.

Comparison of Calculated with Observed Velocities.—The following table, taken, with the exception of the last two columns, from "Annexe à la Table de Krupp," etc., Essen, 1881, shows the agreement between the observed and calculated velocities for projectiles having ogives of 2 calibers. The sixth column gives the distances, in metres, between the points at which the velocities were measured (X_1 and X_2). The seventh and eighth columns give the observed velocities at the distances from the gun X_1 and X_2 respectively. The ninth column gives the velocities at the distances X_2 from the gun computed by Krupp's table and formula. The tenth column gives the velocities at the distances X_2 computed by equation (30), using Table I. of this work. The coefficient of reduction (c) was taken at 0.907, which is its mean value for velocities between 2300 f. s. and 1200 f. s., as determined by a comparison of Bashforth's and Krupp's tables of resistances given in Chapters I. and II. The only discrepancies of any account between the calculated velocities in this column and the observed velocities occur when the curvature of the trajectory is considerable,

No. of Round.	Caliber in mm.	Length of Projectile in Calibers.	Weight in Kilogrammes.	Weight of a Cubic Metre of Air (δ) in Kilogrammes.	$X_1 - X_2$ in Metres. (s)	Measured Velocity at X_1 in Metres. (V)	Measured Velocity at X_2 in Metres. (v)	v Computed by Krupp. m.s.	v Computed by Table I. m.s. c = 0.907.	v Computed by Mayevski's Formulas. m.s.
1	240	2.8	125	1.245	1450	467	380	379.9	380.7	380.6
2	240	2.8	161	1.245	1450	454.5	390	388.3	387.7	387.5
3	172.6	2.8	61.5	1.226	1389	477	388	388.7	389.3	388.7
4	172.6	2.8	61.5	1.226	1429	514.7	416.6	417.9	417.6	415.7
5	149.1	2.8	39.3	1.260	1429	518	401.6	402.1	403.0	401.2
6	149.1	2.5	33.5	1.240	1429	507.7	380	380.7	379.9	379.1
7	149.1	2.8	31.3	1.265	924	475.8	387.8	388.2	387.7	387.3
8	355	2.8	525	1.200	1884	495.9	432.7	433.1	433.8	432.6
9	355	2.8	525	1.200	2384	490	415	411.8	414.4	412.3
10	355	2.8	525	1.200	2389	488.5	409.6	410.4	412.3	410.9
11	149.1	2.8	31.3	1.265	1950	600	394	393.9	395.4	392.7
12	149.1	4	51	1.206	1929	505.2	394.6	393.3	393.4	392.3
13	152.4	4	51.5	1.205	1450	472.4	391.3	389.3	389.1	388.6
14	152.4	2.8	32.5	1.205	1450	577	422	422.0	424.2	421.5
15	149.1	2.8	31.3	1.230	1450	632.4	460.9	460.3	462.8	459.8
16	240	3.8	215	1.208	1904	480.4	412.8	412.0	412.4	411.1
17	400	2.8	777	1.180	2384	499.4	433.7	432.1	433.0	431.7
18	400	2.8	643	1.190	2384	533.4	443.8	447.0	448.2	446.6
19	400	2.8	643	1.190	2384	531.5	444.5	445.4	446.6	445.0
20	84	2.8	6.55	1.197	2447	446.9	266	267.2	259.7	267.4
21	120	2.8	16.4	1.211	2447	463.3	284.1	289.2	281.6	289.3
22	149.1	2.8	31.3	1.285	3448	536.6	294.8	290.6	283.7	290.5
23	105	3.5	16	1.300	3436	481.5	282	278.4	271.2	279.6
24	96	3.5	12	1.340	3439	425.8	256.2	250.5	244.1	254.4
25	107	2.7	12.5	1.218	777.5	205.1	188.2	189.8	187.7	189.8
26	152.4	2.8	31.5	1.206	966.5	203	188	187.4	185.9	188.0
27	105	3.5	16	1.222	950	514.2	426.9	421.1	422.2	420.4
28	149.1	2.8	39	1.218	1429	470	369.5	370.4	369.1	369.3
29	283	2.5	234.7	1.206	4450	464.7	321.2	318.9	311.3	317.6
30	283	2.5	234.7	1.205	1879	465.3	403.9	403.3	404.6	403.7
31	283	2.5	234.7	1.200	1919	465.9	385.4	384.7	384.0	383.8
32	283	2.5	234.7	1.200	2425.5	466.5	370.6	368.0	366.6	367.0
33	283	2.5	234.7	1.220	2921.5	464.8	347.8	350.9	347.7	349.7
34	283	2.5	234.7	1.227	3426.0	463.7	336.0	337.6	331.4	336.6
35	283	2.5	234.7	1.220	4446.5	460.0	316.6	316.6	308.6	315.0
36	283	2.5	234.7	1.192	5945.0	455.8	295.0	293.9	285.6	293.0
37	283	2.5	234.7	1.206	5945.0	453.1	294.7	291.5	283.2	291.4

as in the last four rounds, and one or two others. Equation (30) is based upon the supposition that the path of the projectile is a horizontal right line, and, of course, gives only approximate results when this path has any appreciable curvature. It will be shown subsequently that, to obtain the real velocity, the "v" computed by (30) should be multiplied by the ratio of the cosines of the angles of projection and fall. In No. 37, for example, it will be found that to attain a range of 5945 metres ($3\frac{2}{3}$ miles) the angle of projection would have to be 12° 37′, and the angle of fall would be 17° 40′. Making the necessary correction, we should find the velocity to be 290.7 m.

The last column gives the remaining velocities computed by Mayevski's formulas. They follow very closely those computed by Krupp.

In the absence of tables we may determine remaining velocities which exceed 1300 f. s. as follows: We have found, when $n = 2$,

$$s = \frac{C}{A} \log \frac{V}{v}$$

$$\therefore \frac{V}{v} = e^{\frac{As}{C}} = 1 + \frac{As}{C} + \frac{1}{2}\left(\frac{As}{C}\right)^2 + \text{etc.}$$

As $\frac{As}{C}$ is usually a small quantity, all its powers higher than the first may be neglected, and we may put

$$\frac{V}{v} = 1 + \frac{As}{C}$$

$$\therefore v = \frac{V}{1 + \frac{As}{C}}$$

For oblong projectiles having ogival heads of $1\frac{1}{2}$ calibers $A = 0.000142$. If the ogive is of 2 calibers, $A = 0.0001316$. This method gives correct results for distances of a mile, or even more, especially for the heavy projectiles used with modern seacoast guns. If the data are given in French units —that is w, δ, and δ, in kilogrammes, d in centimetres, and s and V in metres—the value of A will be 0.000030357.

Example. Let $d = 30.5$ cm., $w = 455$ kg., $\delta = 1.274$ kg., $\delta_{,} = 1.206$ kg., $V = 520.8$ m., and $s = 1900$ m.   [Krupp's Bulletin, No. 31.]

We have

$$C = \frac{455 \times 1.206}{(30.5)^2 \times 1.274} = 0.46301$$

and

$$v = \frac{520.8}{1 + \dfrac{0.000030357 \times 1900}{0.46301}} = \frac{520.8}{1.12457} = 463.1 \text{ m.}$$

The measured velocity in this example was 465.5 m., while the velocity computed by Krupp was 460.1 m.

CHAPTER V.

RELATION BETWEEN VELOCITY AND INCLINATION.

Expressions for the Velocity.—Equation (15), which, since $(i) = \dfrac{k^n}{U^n} + (\varphi)$, may be written

$$(\varphi)_n - (\vartheta)_n = k^n \left\{ \frac{1}{u^n} - \frac{1}{U^n} \right\} \tag{31}$$

gives the relation between the horizontal velocities U and u and the corresponding inclinations φ and ϑ; and of these four quantities any three being given, the fourth can be accurately computed, provided, of course, that the value of k has been accurately determined by experiment. The functions $(\varphi)_n$ and $(\vartheta)_n$ are the integrals of $\dfrac{d\vartheta}{\cos^{n+1}\vartheta}$, and the following are the forms they take for the values of n here adopted:

$$(\vartheta)_2 = \tfrac{1}{2} \left\{ \tan \vartheta \sec \vartheta + \log \tan \left(\frac{\pi}{4} + \frac{\vartheta}{2} \right) \right\}$$

$$(\vartheta)_3 = \tan \vartheta + \tfrac{1}{3} \tan^3 \vartheta$$

$$(\vartheta)_6 = \tan \vartheta \left\{ \frac{\sec^6 \vartheta}{6} + \frac{5 \sec^3 \vartheta}{24} + \frac{5 \sec \vartheta}{16} \right\}$$

$$+ \frac{5}{16} \log \tan \left(\frac{\pi}{4} + \frac{\vartheta}{2} \right)$$

It is evident that all these expressions become o when $\vartheta = 0$, negative when ϑ is negative, and infinite when $\vartheta = \dfrac{\pi}{2}$; or, in symbols, $(0) = 0, (-\vartheta) = -(\vartheta)$, and $\left(\dfrac{\pi}{2} \right) = \infty$

If there were but *one* "law of resistance"—in other words, if n had but one value for all velocities—it would be easy to calculate the velocity for any given value of ϑ by means of

(31).   It would only be necessary to tabulate the values of $(\vartheta)_n$
for all practical values of ϑ as the argument, and to pro-
vide a similar table of $\left(\dfrac{k}{u}\right)^n$ with u as the argument.   But, as
we have seen, n may change its value two or three times in
the same trajectory ; and though it would be possible to
ascertain by trial the exact point of the trajectory where
this change occurred, yet the labor involved would be very
great.

Bashforth's Method.—Professor Bashforth overcomes
this difficulty by giving to n the constant value 3, and
making k^3 to vary in such a manner as to satisfy (31) for all
velocities.   His method of procedure is as follows: making
$n = 3$ and $\vartheta = 0$, (31) becomes

$$\frac{1}{v_0^3} - \frac{1}{U^3} = \frac{1}{k^3}\left(\tan\varphi + \frac{1}{3}\tan^3\varphi\right)$$

in which U and φ are the horizontal velocity and inclination,
respectively, at the beginning of any arc of the trajectory
we may be considering ; and v_0 the velocity at the summit.

In Bashforth's notation

$$\frac{1}{k^3} = \frac{3K}{g(1000)^3}\frac{d^2}{w};$$

substituting this in the above equation and multiplying by
$(1000)^3$ to avoid the inconvenience of very small numbers,
we have

$$\left(\frac{1000}{v_0}\right)^3 - \left(\frac{1000}{U}\right)^3 = \frac{K}{g}\frac{d^2}{w}\left\{3\tan\varphi + \tan^3\varphi\right\}$$

by means of which either v_0, U, or φ can be determined
when the other two are known.   When the resistance can
be taken proportional to the cube of the velocity, K is constant; but for all other velocities it is a variable, and we
must take a certain mean of its values for the arc under consideration.   Prof. Bashforth takes the arithmetical mean,
which will generally give very accurate results for arcs of

10 or 15 degrees in extent. In his work he gives the necessary tables for suitably determining $\dfrac{K}{g}$ for all velocities from 100 f. s. to 2900 f. s., and also tables giving values of $3 \tan \varphi + \tan^3 \varphi$ for all practical values of φ.

Other approximate methods involving less labor will be given further on.

High Angle and Curved Fire.—When the initial velocity does not exceed 800 f. s., which includes nearly all mortar and howitzer practice, the law of resistance for oblong projectiles is that of the square of the velocity; whence, making $n = 2$, and dropping the subscript, (31) becomes

$$(\varphi) - (\vartheta) = k^2 \left(\frac{1}{u^2} - \frac{1}{U^2} \right) = \frac{C}{2} \left(\frac{g}{A \, u^2} - \frac{g}{A \, U^2} \right)$$

or, writing $I(u)$ for $\dfrac{g}{A \, u^2}$,

$$(\varphi) - (\vartheta) = \frac{C}{2} \left\{ I(u) - I(U) \right\} \qquad (32)$$

The value of $I(u)$ for any given value of u can be taken directly from Tables I. and II., the method of construction of which will be given further on. Table III. gives (ϑ) and extends from $\vartheta = 0$ to $\vartheta = 60°$.

To use (32) for computing low velocities (and also for high velocities, exceeding 1330 f. s.), we have

$$I(u) = \frac{2}{C} \left\{ (\varphi) - (\vartheta) \right\} + I(U) \qquad (33)$$

in which u and ϑ are the only variables; $\dfrac{2}{C}$, (φ), and $I(U)$, having been determined, do not change their values for the same trajectory.

To illustrate the ease with which velocities may be calculated by (33), take the following data from Bashforth's "Treatise," page 115:

$V = 751$ f. s.; $\varphi = 30°$; $w = 70$ lbs., and $d = 6.27$ inches. Here $U = 751 \cos 30° = 650.385$ f. s.; and from Table I., $I(U) = 0.93354$; $\dfrac{2}{C} = \dfrac{2\,d^2}{w} = 1.12323$.

We will, following Bashforth, compute the velocities for $\vartheta = 28°$, $24°$, $20° \ldots -40°$. The work may be conveniently arranged as follows:

$$(\varphi) = 0.60799 \qquad I(U) = 0.93354.$$

θ	(θ)	$(\phi) - (\theta)$	$\dfrac{2}{C}\big((\phi) - (\theta)\big)$	$I(u)$	(Table I.) u	$u \sec \theta = v$	Bash-forth's v	Differ-ence.
30°	0.60799	0.00000	0.00000	0.93354	650.38	751.0	751.0	0.0
28°	.55580	.05219	.05862	0.99216	636.09	720.4	720.4	0.0
24°	.45953	.14846	.16675	1.10029	612.03	669.5	670.2	− .7
20°	.37185	.23614	.26524	1.19878	592.33	630.3	630.5	.2
16°	.29063	.31736	.35647	1.29001	575.69	598.9	598.9	0.0
12°	.21415	.39384	.44237	1.37591	561.23	573.8	573.5	+ .3
8°	.14100	.46699	.52454	1.45808	548.38	553.8	553.1	.7
4°	.06998	.53801	.60431	1.53785	536.71	538.0	537.0	1.0
0°	.00000	.60799	.68291	1.61645	525.91	525.9	524.6	1.3
−4°	−.06998	.67797	.76151	1.69505	515.74	517.0	515.5	1.5
8°	.14100	.74899	.84129	1.77483	505.99	511.0	509.3	1.7
12°	.21415	.82214	.92345	1.85699	496.52	507.6	505.7	1.9
16°	.29063	.89862	1.00935	1.94289	487.15	506.8	504.7	2.1
20°	.37185	.97984	1.10056	2.03410	477.77	508.4	506.2	2.2
24°	.45953	1.06752	1.19906	2.13260	468.22	512.5	510.2	2.3
28°	.55580	1.16379	1.30720	2.24074	458.38	519.1	516.8	2.3
32°	.66343	1.27142	1.42809	2.36163	448.06	528.3	525.9	2.4
36°	.78617	1.39416	1.56596	2.49950	437.11	540.3	537.9	2.4
40°	.92914	1.53713	1.72654	2.66008	425.32	555.2	552.8	2.4

The numbers in the second column are taken directly from Table III. for the values of ϑ given in column 1. Subtracting the numbers in column 2 from (φ) $(= 0.60799)$ gives those in column 3; and these multiplied by $\dfrac{2}{C}$ $(= 1.12323)$ are written in column 4. Adding $I(U)$ $(= 0.93354)$ to these last gives the values of $I(u)$ in column 5.

The values of u are then taken from Table I., and these multiplied by $\sec \vartheta$ give the velocities sought. For comparison the velocities computed by Bashforth, by his method already explained, are also given; and it will be seen that the differences between his velocities and those computed by (33) are practically *nil.*

This method of determining velocities may be used without material error when the initial velocity is as great as 1000 f. s.

Example.—The 8-inch howitzer is fired with a quadrant elevation of 23°; muzzle velocity, 920 f. s.; weight of shell, 180 lbs.; diameter, 8 inches. What will be the velocity in the descending branch when $\vartheta = -27° \, 54'$? (See Mackinlay's "Text-Book," page 109.)

Here

$$V = 920, \quad U = 920 \cos 23° = 846.86$$

$$I(U) = 0.40884; \quad \log \frac{2}{C} = 9.85194$$

The computation is as follows:

$$
\begin{aligned}
(23°) &= 0.43690 \\
(-27° \, 54') &= -0.55327 \\
\hline
\log 0.99017 &= 9.99571 \\
\log \frac{C}{2} &= 9.85194 \\
\hline
\log 0.70412 &= 9.84765 \\
I(U) &= 0.40884 \\
\hline
I(u) &= 1.11296 \quad \therefore \; U_{27° \, 54'} = 609.4 \text{ f. s.}
\end{aligned}
$$

Mackinlay gets by Niven's method, dividing the trajectory into two parts, $U_{27° \, 54'} = 610.6$ f. s. It will be seen that by the method developed above for calculating velocities, the length of the arc taken makes no difference in the accuracy of the results.

Siacci's Method.—Equation (13) may be written

$$\int_{\theta}^{\phi} \frac{d\vartheta}{\cos^2 \vartheta} = \frac{gC}{A} \int_{u}^{U} \frac{\sec^2 \vartheta \, du}{(u \sec \vartheta)^{n+1}}$$

Since ϑ is a function of u, there must be some constant mean value of sec ϑ which will satisfy the above definite integral. Representing this mean value of sec ϑ by a, and writing U' and u' for aU and au respectively, we have

$$\int_{\theta}^{\phi} \frac{d\vartheta}{\cos^2 \vartheta} = \frac{a g C}{A} \int_{u'}^{U'} \frac{du'}{u'^{n+1}}$$

whence

$$\tan \varphi - \tan \vartheta = a\, k^n \cdot \left(\frac{1}{u'^n} - \frac{1}{U'^n} \right) = \frac{a g C}{n A} \left\{ \frac{1}{u'^n} - \frac{1}{U'^n} \right\} \quad (34)$$

Making

$$I(u') = \frac{2g}{n A} \frac{1}{u'^n} + Q$$

(34) becomes

$$\tan \varphi - \tan \vartheta = \frac{a C}{2} \left\{ I(u') - I(U') \right\} \quad (35)$$

The values of $I(u')$ are given in Table I. for oblong projectiles, and in Table II. for spherical projectiles. The method of computing the I-function is entirely similar to that already described for the S and T-functions, and need not be repeated. For oblong projectiles the formulæ are as follows, in which, for uniformity, $I(v)$ is employed as the general functional symbol:

2800 f. s. $> v >$ 1330 f. s.:

$$I(v) = [5.3547876] \frac{1}{v^3} - 0.028872$$

1330 f. s. $> v >$ 1120 f. s.:

$$I(v) = [8.2947896] \frac{1}{v^3} + 0.015293$$

1120 f. s. $> v >$ 990 f. s.:

$$I(v) = [17.1436868] \frac{1}{v^6} + 0.085087$$

990 f. s. $> v >$ 790 f. s.:

$$I(v) = [8.4557375] \frac{1}{v^3} - 0.061373$$

790 f. s. $> v >$ 0 :

$$I(v) = [5.7369333] \frac{1}{v^3} - 0.356474$$

9

If we compare (34) with (31) it will be seen that

$$a = \left\{ \frac{(\varphi)_n - (\vartheta)_n}{\tan \varphi - \tan \vartheta} \right\}^{\frac{1}{n-1}}$$

and this value of a renders (34) and (35) exact equations; in fact, reduces them to (31). It would seem at first as if nothing had been gained by introducing a into (35), since its value depends upon that of n, and must, therefore, change when n changes. The following table gives the values of a for the arcs contained in the first column, when $n = 2, n = 3$, and $n = 6$, computed by the above formula :

Arc ϕ to θ	$n = 2$ a	$n = 3$ a	$n = 6$ a
30° to 20°	1.1066	1.1069	1.1079
30° " 10°	1.0741	1.0749	1.0772
30° " 0°	1.0531	1.0541	1.0573
30° " −10°	1.0419	1.0429	1.0460
30° " −20°	1.0409	1.0418	1.0443
30° " −30°	1.0531	1.0541	1.0573

It is evident from this table that when the angle of projection is as great as 30°, the velocity at any point of the trajectory may be computed with sufficient accuracy by using either set of values a; since the greatest difference between those in the second and fourth columns on the same line is but 0.0042, and this would make but a slight difference in the values of U' or u'. Moreover, since $U' = a V \cos \varphi$, and $u' = a v \cos \vartheta$, it is apparent that U' and u' differ less from V and v respectively than do U and u; and this is important when, as is usually the case, the law of resistance is different for the initial and terminal velocities.

If in the above expression for a we make $n = 2$, we have Didion's expression for a, viz. :

$$a = \frac{(\varphi) - (\vartheta)}{\tan \varphi - \tan \vartheta}$$

in which

$$(\vartheta) = \frac{1}{2}\left\{ \tan\vartheta \sec\vartheta + \log\tan\left(\frac{\pi}{4} + \frac{\vartheta}{2}\right) \right\}$$

Example.—A 12-inch service projectile, weighing 800 lbs., is fired at an angle of projection of 30° and a muzzle velocity of 1886 f. s. Required its velocity when (a) the inclination of the trajectory is 15°, and (b) when the inclination is — 15°.

Here $d = 12$, $w = 800$, $V = 1886$, and $\varphi = 30°$. From (35) we get

$$I(u') = I(U') + \frac{2}{aC}\left\{ \tan\varphi - \tan\vartheta \right\}$$

(a) $\vartheta = 15°$. From our data we have

$$a = \frac{(30°) - (15°)}{\tan 30° - \tan 15°} = \frac{0.33687}{0.30940} = 1.0888$$

$$U' = a\,V\cos 30° = 1778.34 \quad \therefore I(U') = 0.04270$$

$$C = \frac{w}{d^2} = \frac{800}{144}$$

and

$$\tan 30° - \tan 15° = 0.30940$$

$$\therefore I(u') = 0.04270 + \frac{288 \times 0.30940}{800 \times 1.0888} = 0.14500$$

$$\therefore u' = 1149.77.$$

$$\therefore v_{15°} = \frac{1149.77}{a\cos 15°} = 1093.3 \text{ f. s.}$$

(b) $\vartheta = -15°$. We have

$$a = \frac{(30°) + (15°)}{\tan 30° + \tan 15°} = \frac{0.87911}{0.84530} = 1.0400$$

$$U' = a\,V\cos 30° = 1698.65 \quad \therefore I(U') = 0.04958$$

$$\tan 30° + \tan 15° = 0.84530$$

$$\therefore I(u') = 0.04958 + 0.29260 = 0.34218$$

$$\therefore u' = 891.14$$

$$\therefore v_{-15°} = 887.1 \text{ f. s.}$$

The values of $v_{15°}$ and $v_{-15°}$ computed by (31) are 1097.6 and 892.9 respectively.

Siacci's Modification of (35) for Direct Fire.—

Since in direct fire the angle of projection does not exceed 15°, and is generally much less, the values of a for this kind of fire will not differ much from unity. For example, with 10° elevation, and an angle of fall of — 12°, we shall have for a

$$a = \frac{(10°) + (12°)}{\tan 10° + \tan 12°} = \frac{0.39139}{0.38889} = 1.0064$$

It is manifest, therefore, that for such small angles no material error would result in making $a = 1$; the following, however, is a closer approximation. If we consider that part of the trajectory lying above the horizontal plane passing through the muzzle of the gun, it will be seen that a should be greater than unity and less than sec ω. Siacci makes

$$a = (\sec \varphi)^{\frac{n-2}{n-1}}$$

therefore, when $n = 2$, $a = 1$; when $n = 3$, $a = \sqrt{\sec \varphi}$, and when $n = 6$, $a = \sec^4 \varphi$; and the average value of a for the whole trajectory generally fulfils the above condition.

This value of a substituted in (34) gives, by an easy reduction,

$$\tan \varphi - \tan \vartheta = \frac{g\,C}{n\,A\,\cos^2 \varphi} \left\{ \frac{1}{(u \sec \varphi)^n} - \frac{1}{V^n} \right\}$$

or, writing u' for $u \sec \varphi$, and proceeding as already explained,

$$\tan \varphi - \tan \vartheta = \frac{C}{2 \cos^2 \varphi} \left\{ I(u') - I(V) \right\} \qquad (36)$$

Example.—Take the following data:

$d = 12$; $w = 800$; $C = \dfrac{800}{144}$; $V = 1886$; $\varphi = 10°$. Compute the remaining velocity in the descending branch when $\vartheta = - 13°$. We have

$$I(u') = \frac{2}{C} \cos^2 \varphi (\tan \varphi - \tan \vartheta) + I(V)$$

and the computation will be as follows:

$$\log (\tan 10° + \tan 13°) = 9.60980$$

$$\log \frac{2}{C} = 9.55630$$

$$2 \log \cos 10° = 9.98670$$

$$\log 0.14217 = 9.15280$$

$$I (1886) = 0.03477$$

$$I (u') = 0.17694 \qquad u' = 1071.76$$

$$\therefore v = \frac{1071.76 \cos 10°}{\cos 13°} = 1083.2 \text{ f. s.}$$

The velocity at the same point computed by (31), dividing the trajectory into three arcs, with the points of division corresponding to velocities of 1330 f. s. and 1120 f. s. respectively, is $v = 1081.55$ f. s. This agreement is very close; but if we make $\varphi = 30°$ and $\vartheta = 15°$, as in the preceding example, we should find by this method $v_{15°} = 1113.1$; and if $\vartheta = -15°$, we should find $v_{-15°} = 859.3$, which differ considerably from their true values.

Niven's Method.—W. D. Niven, Esq., M.A., F.R.S., has given the following method for determining velocities in terms of the inclination:

Equation (13) may be written

$$\int_\theta^\phi d\vartheta = \frac{G C}{A} \int_u^v \frac{du}{(u \sec \vartheta)^{n+1}} = \frac{g C}{a A} \int_{u'}^{v'} \frac{du'}{u'^{n+1}}$$

in which, as before, a is some mean value of $\sec \vartheta$ between the limits $\sec \varphi$ and $\sec \vartheta$, and $u' = a v \cos \vartheta$ and $U' = a V \cos \varphi$. Integrating, we have

$$\varphi - \vartheta = \frac{g C}{a n A} \left\{ \frac{1}{u'^n} - \frac{1}{U'^n} \right\} = \frac{C}{a} \left\{ \frac{g}{n A} \frac{1}{u'^n} - \frac{g}{n A} \frac{1}{U'^n} \right\} \quad (37)$$

Multiplying both members by $\frac{180}{\pi}$ to reduce $\varphi - \vartheta$ to degrees, and making

$$\frac{180}{\pi} (\varphi - \vartheta) = D$$

and

$$\frac{180\,g}{n\,\pi\,A}\,\frac{1}{n'^n} = (D\,u')$$

the above equation becomes

$$D = \frac{C}{a}\left\{ D\,(u') - D\,(U') \right\}^{*} \qquad (38)$$

which is the equivalent of Niven's expression for the velocity and inclination. Mr. Niven has published a table of the D-function for velocities extending from 400 f. s. to 2500 f. s. (See Table VI. in Mackinlay's "Text-Book.") It will be seen by comparing the expressions for $D\,(v)$ and $I\,(v)$ that we have the relation

$$D\,(v) = \frac{90}{\pi}\,I\,(v)$$

and, therefore, in terms of the I-function, (38) becomes

$$D = \frac{90\,C}{a\,\pi}\left\{ I\,(u') - I\,(U') \right\} \qquad (39)$$

$$\log \frac{90}{\pi} = 1.4570926$$

Comparing (37) with (31), it is apparent that to make (37) or (38) exact equations we must have

$$a = \left\{ \frac{(\varphi)_n - (\vartheta)_n}{\varphi - \vartheta} \right\}^{\frac{1}{n+1}}$$

For direct fire Didion's value of a may be used; but for high-angle firing the following gives more accurate results, obtained from the above equation by making $n = 2$:

$$a = \left\{ \frac{(\varphi) - (\vartheta)}{\varphi - \vartheta} \right\}^{\frac{1}{3}}$$

Example.—Take the following data:

$d = 12$; $w = 800$; $V = 1886$; $\varphi = 30°$ and $\vartheta = -30°$; $D = 30° + 30° = 60°$; to find $v_{-30°}$.

---

* If we use Niven's tables, in which the functions *decrease* with the velocity, (38) should be written

$$D = \frac{C}{a}\left\{ D\,(U') - D\,(u') \right\}$$

We have from (38)

$$D(u') = D(U') + \frac{a}{C} D$$

The computation may be conveniently arranged as follows:

log $(\varphi) = 9.78390$
constant $= 1.75812$
c log $30 = 8.52288$

3)$\overline{0.06490}$

log $a = 0.02163$	log $V = 3.27554$
log $D = 1.77815$	log $a = 0.02163$
c log $C = 9.25527$	log cos $\varphi = 9.93753$
log $11.3516 = 1.05505$	log $U' = 3.23470$
	$U' = 1716.74$

$$\text{(Niven's Table)} \quad D(U') = 84.6090$$

$$\frac{a}{C} D = 11.3516$$

$$D(u') = 73.2574$$

$$\therefore u' = 827.12 = a\, v \cos \vartheta$$
$$\therefore v_{-30°} = 908.7 \text{ f. s.}$$

Siacci's method, using Table I. of this work, gives $v_{-30°} = 907.5$ f. s.; while equation (31) gives $v_{-30°} = 913.2$ f. s.
Modification of (38) for Direct Fire.—If we make

$$a = (\sec \varphi)^{\frac{n-1}{n}}$$

we shall have, by a process similar to that already employed in Siacci's method, the following modified form of (38), which can be used in all problems of direct fire, viz.:

$$D = \frac{C}{\cos \varphi} \left\{ D(u') - D(V) \right\} \qquad (40)$$

in which $u' = u \sec \varphi$.

Example.—Let $d = 12$; $w = 800$; $V = 1886$; $\varphi = 10°$; $\vartheta = -13°$. The computation is as follows:

$$\log D = 1.36173$$
$$\log \cos \varphi = 9.99335$$
$$c \log C = 9.25527$$
$$\log 4.0771 = \overline{0.61035}$$
$$D\,(1886) = 84.9966$$
$$D\,(u') = \overline{80.9195} \quad \therefore\ u' = 1068.14 = v\frac{\cos \vartheta}{\cos \varphi}$$
$$\therefore\ v = 1079.6 \text{ f. s.}$$

which is within 2 feet of the value of v computed by the exact formula (31). This modified form of Niven's method, for simplicity and accuracy, seems to leave nothing to be desired.

For small angles of projection, say not exceeding 5°, we may put $u' = v$, and $\cos'\varphi = 1$; and (40) becomes

$$D = C\left\{ D\,(v) - D\,(V) \right\} = \frac{90}{\pi} C\left\{ I\,(v) - I\,(V) \right\}$$

, *Example.*—In the preceding example suppose $\varphi = 3°$. What will be the value of ϑ when the velocity is reduced to 1500 f. s.?

(a) By Niven's Table:
$$D\,(1886) = 84.9966$$
$$D\,(1500) = \overline{83.9359}$$
$$\log 1.0607 = 0.02560$$
$$\log C = 0.74473$$
$$\log D = \overline{0.77033}$$
$$D = 5°.89 = 3° - \vartheta$$
$$\therefore\ \vartheta = -2°.89$$

(b) By Table I.:
$$I\,(1500) = 0.07173$$
$$I\,(1886) = \overline{0.03477}$$
$$\log 0.03696 = 8.56773$$
$$\log \frac{90}{\pi} = 1.45709$$
$$\log C = 0.74473$$
$$\log D = \overline{0.76955}$$
$$D = 5°.88$$
$$\therefore\ \vartheta = -2°.88$$

CHAPTER VI.

TRAJECTORIES—HIGH-ANGLE FIRE.

As we have seen, the differential equations for $x, y,\ t$, and s do not generally admit of integration in finite terms for any law of resistance pertaining to our atmosphere; that is, for any recognized value of n. It is true that Professor Greenhill has recently * succeeded, by a profound analysis, in deducing exact finite expressions for x and y by means of elliptic functions, when $n = 3$; but these results, though of great interest to the mathematician, are far too complicated for the practical use of the artillerist. When $n = 2$ the expression for ds can be integrated and useful results deduced therefrom, as will be seen further on.

For low velocities, such as are generally employed in high-angle and curved fire, the effect of the resistance of the air upon heavy projectiles is comparatively slight; and for a first (though rough) approximation we may, in such cases, omit the resistance altogether, or, better still, we may suppose the projectile subject to a *mean constant* resistance. A still closer approximation may be obtained by taking a resistance proportional to the first power of the velocity. As the differential equations for the co-ordinates and time are susceptible of exact integration upon each one of these hypotheses, we will consider them in turn.

TRAJECTORY IN VACUO.

Making $\rho = 0$, (9) becomes

$$du = 0$$

and therefore, *in vacuo*, the horizontal velocity is constant, or

$$u = U$$

Integrating (21), (22), (23), and (24) between the limits φ and ϑ gives, if $u = U$,

---

* " Proceedings of the Royal Artillery Institution," Vol. XI.

$$t = \frac{U}{g}(\tan \varphi - \tan \vartheta) \tag{41}$$

$$x = \frac{U^2}{g}(\tan \varphi - \tan \vartheta) \tag{42}$$

$$y = \frac{U^2}{2g}(\tan^2 \varphi - \tan^2 \vartheta) \tag{43}$$

and

$$s = \frac{U^2}{g}\left((\varphi) - (\vartheta)\right) \tag{44}$$

Equation of Trajectory in Vacuo.—Eliminating $\tan \vartheta$ from (42) and (43) gives

$$y = x \tan \varphi - \frac{g x^2}{2 U^2}$$

which is the equation of a parabola whose axis is vertical A parabola, therefore, is the curve a projectile would describe *in vacuo*.

Since a parabola is symmetrical with respect to its axis, the ascending branch is similar in every respect to the descending branch, the angle of fall being equal to the angle of projection; and generally, for the same value of y, $\tan \vartheta$ has numerically the same value, but with contrary signs, in both branches; being positive in the ascending branch, negative in the descending branch, and zero at the vertex.

If we make $\vartheta = -\varphi$ in (42) it becomes

$$X = \frac{2 U^2}{g} \tan \varphi = \frac{V^2 \sin 2\varphi}{g}$$

and this, for a given velocity, is evidently a maximum when $\varphi = 45°$.

Subtracting (42) from the above equation, and reducing, gives

$$X - x = \frac{X}{2 \tan \varphi}(\tan \varphi + \tan \vartheta)$$

also, dividing (43) by (42) gives

$$\frac{y}{x} = \frac{1}{2}(\tan \varphi + \tan \vartheta)$$

whence

$$y = \frac{x}{X}(X - x) \tan \varphi \tag{45}$$

Making $\vartheta = -\varphi$ in (41), we have

$$T = \frac{2U}{g} \tan \varphi = \frac{2V}{g} \sin \varphi$$

Subtracting (41) from this last equation gives

$$T - t = \frac{U}{g} (\tan \varphi + \tan \vartheta)$$

also, (43) divided by (41) gives

$$\frac{y}{t} = \frac{U}{2} (\tan \varphi + \tan \vartheta)$$

whence

$$y = \frac{gt}{2} (T - t) \qquad (46)$$

Dividing (44) by (42) gives

$$\frac{s}{x} = \frac{(\varphi) - (\vartheta)}{\tan \varphi - \tan \vartheta} = a$$

Didion's a, then, is the ratio of a parabolic arc whose extremities have the same inclination as the arc of the trajectory under consideration, to its horizontal projection.

Expression for the Velocity.—From (43) we have, since $V \cos \varphi = v \cos \vartheta = U$,

$$v^2 \sin^2 \vartheta = V^2 \sin^2 \varphi - 2 g y.$$

Adding $v^2 \cos^2 \vartheta$ to the first member, and its equal, $V^2 \cos^2 \varphi$, to the second member, and reducing, we have

$$v^2 = V^2 - 2 g y$$

If h is the vertical height through which the projectile must fall to acquire the velocity of projection (V), we shall have

$$V^2 = 2 g h$$

and this substituted in the above formula gives

$$v^2 = 2 g (h - y)$$

that is, the velocity of the projectile at any point of the trajectory is that which it would acquire by falling through a vertical distance equal to $h - y$.

All the properties of the trajectory *in vacuo* may be easily and elegantly determined by means of the fundamental equations (41) to (44) inclusive.

CONSTANT RESISTANCE.

Suppose the resistance constant, and put $\dfrac{\rho}{w} = m$; then the elimination of dt from (9) and (12) gives

$$\frac{du}{u} = m \frac{d\vartheta}{\cos \vartheta}$$

whence

$$\log u = m \log \tan \left(\frac{\pi}{4} + \frac{\vartheta}{2} \right) + C. \cdot$$

Let v_0 be the velocity when $\vartheta = 0$, that is, at the summit of the trajectory; then $C = \log v_0$, and we have

$$u = v_0 \tan^m \left(\frac{\pi}{4} + \frac{\vartheta}{2} \right) \qquad (47)$$

Substituting this value of u in equations (21) to (24), and integrating so that t, x, y, and s shall all be zero at the origin, that is, when $\vartheta = \varphi$, we have, making the necessary reductions,

$$t = V \frac{\sin \varphi - m}{g (1 - m^2)} - v \frac{\sin \vartheta - m}{g (1 - m^2)}$$

$$x = V^2 \frac{\cos \varphi (\sin \varphi - 2m)}{g (1 - 4m^2)} - v^2 \frac{\cos \vartheta (\sin \vartheta - 2m)}{g (1 - 4m^2)}$$

$$y = V^2 \frac{1 + \sin \varphi (\sin \varphi - 2m)}{4g (1 - m^2)} - v^2 \frac{1 + \sin \vartheta (\sin \vartheta - 2m)}{4g (1 - m^2)}$$

$$s = V^2 \frac{\cos^2 \varphi + 2m (\sin \varphi - m)}{4mg (1 - m^2)} - v^2 \frac{\cos^2 \vartheta + 2m (\sin \vartheta - m)}{4mg (1 - m^2)}$$

When $2m = 1$, the differential expression for x becomes logarithmic, as do those for t, y, and s when $m = 1$. The integrations are easily obtained for these values of m, but are omitted on account of their length, and as being of no great practical importance. In the application of these formulæ it will be necessary, since the resistance of the air is not constant, but varies with the velocity, to determine a proper mean value for m between the limits of integration; and this we may do as follows: After having computed the horizontal velocities u_α and u_β by means of (33), corresponding to the inclinations α and β, the value of m may be determined by the following equation deduced from the above expression for u:

$$m = \frac{\log u_a - \log u_\beta}{\log \tan \left(\dfrac{\pi}{4} + \dfrac{a}{2}\right) - \log \tan \left(\dfrac{\pi}{4} + \dfrac{\beta}{2}\right)}$$

Example.—Compute the values of t, x, y, and s, from $\varphi = 30°$ to $\vartheta = 0$, with the data given on page 67. We have

$$m = \frac{\log 751 + \log \cos 30° - \log 525.91}{\log \tan 60.°} = 0.38673$$

Substituting in the above formulæ, we find

$$t = 3.1073 + 7.4295 = 10''.537$$
$$x = 16908 - 10557 = 6351 \text{ ft.}$$
$$y = 4446 - 2526 = 1920 \text{ ft.}$$
$$s = 11155 - 4578 = 6577 \text{ ft.}$$

Bashforth gets, by dividing the arc into 8 parts, $t = 10''.413$, $x = 6074$ ft., and $y = 1882$ ft.

It is easy to see how by suitable tables, the construction of which offers no difficulty, the time and co-ordinates may by this method be readily, and for arcs of limited extent accurately, computed. For example, we have

$$x = A V^2 - A' v^2$$

A being a function of m and φ, and A' the same function of m and ϑ.

RESISTANCE PROPORTIONAL TO THE FIRST POWER OF THE VELOCITY.

Differential Equations.—When $n = 1$, the differential equations (13), (17), (18), and (19) become respectively, since $k^n = \dfrac{g C}{n A}$,

$$\frac{d\vartheta}{\cos^2 \vartheta} = k \frac{du}{u^2}$$
$$dt = -\frac{k}{g} \frac{du}{u}$$
$$dx = -\frac{k}{g} du$$
$$dy = -\frac{k}{g} \tan \vartheta \, du$$

Time and Co-ordinates.—The integration of the first three of these equations between the limits (φ, ϑ) and (U, u) gives (supposing k constant)

$$\tan \varphi - \tan \vartheta = k\left(\frac{1}{u} - \frac{1}{U}\right) \qquad (48)$$

$$t = \frac{k}{g} \log \frac{U}{u}$$

or, using common logarithms,

$$t = M \frac{k}{g} \log \frac{U}{u} \qquad (49)$$

in which $M = 2.30259$; and

$$x = \frac{k}{g}(U - u) \qquad (50)$$

Substituting for $\tan \vartheta$ in the expression for dy its value from (48), it becomes

$$dy = -\frac{k}{g}\left(\frac{k}{U} + \tan \varphi\right) du + \frac{k^2}{g}\frac{du}{u}$$

or

$$dy = \left(\frac{k}{U} + \tan \varphi\right) dx - k dt$$

whence, supposing y to vanish with x and t,

$$y = \left(\frac{k}{U} + \tan \varphi\right) x - kt \qquad (51)$$

Determination of k.—In the above integrations we have assumed k to be constant, whereas it varies with the velocity; but our results will be correct if we give to k a proper mean of all its values between the limits of integration; and as k varies slowly and with considerable regularity for all velocities for which this method will be used, we will take for k the value corresponding to the arithmetical mean of the two velocities at the extremities of the arc under consideration. It is evident that the smaller the arc of the trajectory over which we integrate, the less will be the error committed in taking this value for k. But it will be shown by examples that no material error will result for velocities less than about 1000 f. s., when the whole trajectory is divided into two arcs with the point of division at the summit.

When $n = 1$, we have

$$\frac{g}{w}\rho = \frac{g}{k}v$$

whence from (6) and (7)

$$\frac{k}{g} = C\frac{(1000)^3}{Kv^3} = Cm \quad \text{(say)}$$

The following table gives the values of m for velocities extending from 900 f. s. to 500 f. s., with first differences:

TABLE OF m.

v	m	d_i	v	m	d_i
500	32.814	668	710	23.700	346
510	32.146	618	720	23.354	357
520	31.528	572	730	22.997	340
530	30.956	554	740	22.657	323
540	30.402	539	750	22.334	335
550	29.863	527	760	21.999	376
560	29.336	490	770	21.623	388
570	28.846	427	780	21.235	372
580	28.419	392	790	20.863	358
590	28.027	387	800	20.505	344
600	27.640	384	810	20.161	384
610	27.256	381	820	19.777	448
620	26.875	382	830	19.329	433
630	26.493	382	840	18.896	442
640	26.111	356	850	18.454	426
650	25.755	388	860	18.028	412
660	25.367	365	870	17.616	398
670	25.002	343	880	17.218	385
680	24.659	321	890	16.833	372
690	24.338	300	900	16.461	359
700	24.038	338			

The value of k in the ascending branch will be assumed to be that due to the velocity $\frac{1}{2}(V+v_0)$; and in the descending branch, to $\frac{1}{2}(v_0+v_\theta)$, v_θ being the velocity at the point of fall. The first step, then, is to compute v_0 and v_θ; and this can readily be done by means of (33), as already explained.

Expressions for the Ascending and Descending Branches.—It will be seen that x, y, and t are functions of U and u; and these latter depend upon φ and ϑ, as shown in equation (48).

From this equation we have

$$\frac{k}{U} + \tan \varphi = \frac{k}{u_\theta} + \tan \vartheta = \frac{k}{u_0}$$

in which u_0 is the value of u at the summit; whence

$$u_0 = \frac{k}{\dfrac{k}{U} + \tan \varphi} \tag{52}$$

and, since ϑ is negative in the descending branch,

$$u_\theta = \frac{k}{\dfrac{k}{v_0} + \tan \vartheta} \tag{53}$$

The following expressions for t, x, and y for the ascending and descending branches are easily deduced from (49), (50), and (51), in connection with (52) and (53):

ASCENDING BRANCH.	DESCENDING BRANCH.
$t_0 = M \dfrac{k}{g} \log \dfrac{U}{u_0}$	$t_\theta = M \dfrac{k}{g} \log \dfrac{v_0}{u_\theta}$
$x_0 = \dfrac{k}{g}\left(U - u_0\right)$	$x_\theta = \dfrac{k}{g}\left(v_0 - u_\theta\right)$
$y_0 = \dfrac{k}{u_0} x_0 - k\, t_0$	$y_\theta = \dfrac{k}{v_0} x_\theta - k\, t_\theta$

In using these formulæ, u_0 and u_θ are to be computed by means of (52) and (53).

The zero subscript is to be interpreted " from the origin to the summit"; and the theta subscript "from the summit

to a point in the descending branch where the inclination
is ϑ."

The method of computing a trajectory by these simple
formulæ will be best exhibited by examples, which we will
select from those that have been worked out by other
methods of recognized accuracy, or which have been tested
by firing.

Example 1.—Calculate the trajectory with the data on
page 67, viz. :

$V = 751$ f. s.; $\varphi = 30°$ (whence $U = V \cos \varphi = 650.385$); $d =$
6.27 inches; $w = 70$ lbs. (whence $\dfrac{2}{C} = \dfrac{2d^2}{w} = 1.12323$).

Assuming $-37°$ to be the angle of fall, we will divide
the trajectory into two arcs, the first extending from 30° to
0°, and the second from 0° to $-37°$. The velocities v_0 and
$v_{-37°}$ are computed as follows:

From Table III. we take out $(30°) = 0.60799$, and $(37°) =$
0.81977; and from Table I., $I(U) = I(650.385) = 0.93354$.
Then

$$\frac{2}{C}(30°) = 1.12323 \times 0.60799 = 0.68291$$

$$I(U) = 0.93354$$

$$I(v_0) = 1.61645$$

$$\text{(Table I.)} \quad v_0 = 525.91$$

$$\frac{2}{C}(37°) = 1.12323 \times 0.81977 = 0.92079$$

$$I(v_0) = 1.61645$$

$$I(u_{-37°}) = 2.53724$$

$$u_{-37°} = 434.25$$

$$v_{-37°} = 434.25 \sec 37° = 543.74 \text{ f. s.}$$

The mean velocity from which to determine k in the
ascending branch is $\frac{1}{2}(751 + 525.91) = 638$ f. s.; whence
$m = 26.187$. The remaining calculations may be conveniently arranged as follows:

11

$$\log m = 1.4180857$$
$$\log C = 0.2505630$$
$$\log g = 1.5077210 \quad (g = 32.19)$$
$$\overline{\log k = 3.1763697}$$
$$\log U = 2.8131705$$
$$\log 2.3078 = \overline{0.3631992} = \log \frac{k}{U}$$

[Equation (52)] $\tan \varphi = 0.5774$

$$\log 2.8852 = 0.4601759 \,(\text{sub. from } \log k)$$
$$\log u_0 = \overline{2.7161938}$$
$$u_0 = 520.228$$
$$U = 650.385$$
$$\log 130.157 = \overline{2.1144675}$$
$$\log \frac{k}{g} = 1.6686487$$
$$\log x_0 = \overline{3.7831162}$$
$$x_0 = 6069 \text{ ft.}$$

Bashforth gets by 8 steps, $\overline{6074}$

       Difference,   5 ft.

$$\log U = 2.8131705$$
$$\log u_0' = 2.7161938$$
$$\log 0.0969767 = \overline{8.9866674}$$
$$\log M = 0.3622157 \left(\text{add } \log \frac{k}{g} \right)$$
$$\log t_0 = \overline{1.0175318}$$
$$t_0' = 10''.412$$

Bashforth gets $10''.413$

Difference, $\overline{0''.001}$

$$\log \frac{x_0}{u_0} = 1.0669224 \quad (\text{add } \log k)$$
$$\overline{4.2432921} = \log 17510$$
$$\log k t_0 = 4.1939015 = \log 15628$$
$$y_0 = \overline{1882}$$

Bashforth gets   1882

Difference, $\overline{0}$

These results, being practically identical with those deduced with vastly greater labor by Prof. Bashforth, prove that when the law of resistance is that of the square of the velocity, as in this example, we may get quite as close an approximation to the true trajectory by assuming that the resistance is proportional to the first power of the velocity as we can upon the hypothesis of the law of the cube, and with a great gain in simplicity and labor.

We have next to compute the descending branch from $\vartheta = 0°$ to $\vartheta = -37°$. The mean velocity from which to determine k in this branch is

$$\tfrac{1}{2}(525.91 + 543.74) = 534.8 \text{ f. s.}$$

whence $m = 30.690$.

$$\log m = 1.4869969$$
$$\log C = 0.2505630$$
$$\log g = 1.5077210$$

$$\log k = 3.2452809$$
$$\log v_0 = 2.7209114$$

[Equation (53)] $\log 3.34480 = 0.5243695 = \log \dfrac{k}{v_0}$

$$\tan 37° = 0.75355$$

$$\log 4.09835 = 0.6126090$$

$$\log u_{-37°} = 2.6326719$$
$$u_{-37°} = 429.21$$
$$v_0 = 525.91$$

$$\log 96.70 = 1.9854265$$
$$\log \dfrac{k}{g} = 1.7375599$$

$$\log x_{-37°} = 3.7229864$$
$$x_{-37°} = 5284 \text{ ft.}$$

$$\log v_0 = 2.7209114$$
$$\log u_{-37°} = 2.6326719$$

$$\log 0.0882395 = 8.9456631$$
$$\log M = 0.3622157$$

$$\log t_{-37°} = 1.0454387$$
$$t_{-37°} = 11''.103$$

$$\log \frac{k}{v_0} x_{-37°} = 4.2473559 = \log 17675$$

$$\log k\, t_{-37°} = 4.2907196 = \log 19531$$

$$y_{-37°} = -\ \overline{\quad 1856}\ \text{ft.}$$

The projectile is still $1882 - 1856 = 26$ ft. above the level of the gun $= \Delta y$. If Δx and Δt are the corresponding additions to the range and time of flight, we shall have approximately

$$\Delta x = 26 \cot 37° = 35 \text{ ft.; and } \Delta t = \frac{\Delta x}{u_{-37°}} = 0''.080.$$

We therefore have

$$X = 6069 + 5284 + 35 = 11388 \text{ ft.}$$
$$T = 10.412 + 11.103 + 0.080 = 21''.595$$

These values agree almost exactly with those deduced by interpolation from the table on page 117 of Bashforth's work.

Example 2.—The 8-inch howitzer is fired with a quadrant elevation of 23°. Muzzle velocity, 920 f. s.; weight of shell, 180 lbs.; diameter, 8 inches. Find the range and time of flight. (Mackinlay's "Text-Book of Gunnery," page 107.)

Assuming the angle of fall to be $- 27° 54'$, we find by the above method

$$X = 7886 + 7108 - 13 = 14981 \text{ ft.}$$
$$T = 10.183 + 10.801 - 0.022 = 20''.962$$

Mackinlay gets, using Niven's method,

$$X = 14787 \text{ ft., and } T = 20''.813$$

He states that "the published range-table gives 15000 ft. as the range, and $21''.5$ for the time of flight."

Example 3.—Let $V = 298$ m. $= 977.71$ ft., $d = 15$ c.m., $w = 30$ k.g., $\varphi = 35° 21'$, $\delta = 1.270$ k.g., and $\delta_{,} = 1.206$ k.g. Find X and T. (Krupp's Bulletin, No. 55, December, 1884.)

For the Krupp projectiles and low velocities we will take for c the ratio of the coefficients of resistance deduced from the Krupp and Bashforth experiments respectively, and which are given in Chapter II. Let these coefficients be represented by A and A'. Then for velocities less than 790 f. s. we have

$$\log A = 5.6698755 - 10$$
$$\log A' = 5.7703827 - 10$$
$$\log c = 9.8994928$$
$$\therefore c = 0.7934$$

To find C, expressed in English units, when w and d are given in kilogrammes and centimetres respectively, we have

$$C = \frac{10000\, k}{144\, m^2\, c}\, \frac{w}{d^2}$$

in which k is the number of pounds in one kilogramme, and m the number of feet in one metre. Reducing, we have

$$C = [1.2534887]\frac{w}{d^2}$$

As the initial velocity in this example is considerable, we will take into account the density of the air at the time the shots were fired, and also the diminution of density due to the altitude attained by the projectile; and for this purpose we will assume the mean value of y for the whole trajectory to be 2000 ft.

The complete expression for C is (Chapter VII.),

$$C = \frac{w}{d^2}\frac{\delta_{,}}{\delta}e^{\frac{y}{\lambda}}$$

from which we determine $\log C$ as follows:

$$\log w = 1.4771213$$
$$c \log d^2 = 7.6478175$$
$$\text{constant } \log = 1.2534887$$
$$\log \delta_{,} = 0.0813473$$
$$c \log \delta = 9.8961963$$
$$\log e^{\frac{y}{\lambda}} = 0.0312468$$
$$\log C = 0.3872179$$

Assuming the angle of fall to be $-44°\ 40'$, and proceeding as in the first example, we find

$$X = 10408 + 8736 + 104 = 19248 \text{ ft.}$$
$$T = 15.088 + 16.324 + 0.221 = 31''.633$$

Krupp gives the ranges of three shots fired with the initial velocity and angle of departure of this example, and the ranges reduced to the level of the mortar, as follows:

NO. OF SHOT.	RANGE IN FEET.
18	19039
19	19265
20	19364

Mean of the three shots $= 19223$ ft.
Computed—mean $=\quad 25$ ft.

Example 4.—Given $V = 206.6$ m. $= 677.834$ ft., $d = 21$ c.m., $w = 91$ k.g., and $\varphi = 60°$, to find X and T. (Krupp's Bulletin, No. 31, Dec. 30, 1881.)

It will be found that (assuming the angle of fall to be $-63°\ 30'$, and taking no account of atmospheric conditions)

$$X = 5390 + 4945 + 67 = 10402 \text{ ft.}$$
$$T = 17.016 + 17.543 + 0.250 = 34''.809$$

Krupp gives the observed ranges of five shots, with the above data, as follows:

NO. OF SHOT.	OBSERVED RANGE.
22	10332 ft.
23	10305 "
24	10384 "
25	10463 "
26	10440 "

Mean of the five shots $= 10385$ ft.
Computed—mean $=\quad 17$ ft.

Example 5.—Given $V = 204.1$ m. $= 669.63$ ft., $d = 21$ c.m., $w = 91$ k.g., and $\varphi = 45°$, to find X and T. (Krupp's Bulletin, No. 31, January 19, 1882.)

Assuming the angle of fall to be $-49°$, we find as follows:

$$X = 6152 + 5678 + 56 = 11886 \text{ ft.}$$
$$T = 13.817 + 14.238 + 0.147 = 28''.202$$

The following ranges were measured at Meppen:

NO. OF SHOT.	OBSERVED RANGE.
71	11923 ft.
72	11920 "
73	11841 "
74	11808 "
75	11749 "

Mean of the five shots = 11848 ft.
Computed—mean   =   38 ft.

Example 6.—Compute X and T with the data of the pre.
ceding example, except that $\varphi = 30°$.

Assuming the angle of fall to be $-33°$, we find as follows :

$$X = 5478 + 5143 + 26 = 10647 \text{ ft.}$$
$$T = 9.908 + 10.183 + 0.054 = 20''.145$$

Krupp gives as the mean of five measured ranges,
$X = 10779$ ft.
Computed—mean $= -132$ ft.

EULER'S METHOD.

Expression for *s*.—If we make $n = 2$, that is, suppose
the resistance of the air proportional to the square of the
velocity, we shall have from (20)

$$ds = -\frac{C}{A}\frac{du}{u}$$

whence, integrating and supposing $s = 0$ when $u = U$, we
have

$$s = \frac{C}{A}\left\{ \log U - \log u \right\}$$

therefore (page 52)

$$s = C[S(u) - S(U)] \qquad (54)$$

which gives the length of any arc of a trajectory when the
resistance is proportional to the square of the velocity, by
means of the table of space functions.

We may also obtain another expression for *s*, better
suited to our purpose, as follows:

Since

$$(\vartheta) = \int \frac{d\vartheta}{\cos^{n+1}\vartheta}$$

we have, when $n = 2$,

$$d(\vartheta) = \frac{d\vartheta}{\cos^3\vartheta} = \sec\vartheta\, d\tan\vartheta$$

and this substituted in (28) gives

$$ds = -\frac{k^2}{g}\frac{d(\vartheta)}{(\imath) - (\vartheta)}$$

in which

$$(\vartheta) = \tfrac{1}{2}\left\{\tan\vartheta\sec\vartheta + \log\tan\left(\frac{\pi}{4} + \frac{\vartheta}{2}\right)\right\}$$

whence, integrating between the limits φ and ϑ, we have

$$s = \frac{k^2}{g}\log\frac{(\imath) - (\vartheta)}{(\imath) - (\varphi)}$$

or, if we use common logarithms,

$$s = M\frac{k^2}{g}\log\frac{(\imath) - (\vartheta)}{(\imath) - (\varphi)} \qquad (55)$$

in which $M = 2.30259$.

Expressions for x and y.—Equation (55) gives the value of s from the origin. If s' is the length of an arc of the trajectory from the origin to where the inclination is ϑ', and s'' the length to some other point further on where the inclination is ϑ'' ($\vartheta' > \vartheta''$), we shall have from (55)

$$s' = M\frac{k^2}{g}\log\frac{(\imath) - (\vartheta')}{(\imath) - (\varphi)}$$

and

$$s'' = M\frac{k^2}{g}\log\frac{(\imath) - (\vartheta'')}{(\imath) - (\varphi)}$$

whence

$$s'' - s' = \varDelta s = M\frac{k^2}{g}\log\frac{(\imath) - (\vartheta'')}{(\imath) - (\vartheta')}$$

If ϑ'' differs but little from ϑ' (say one degree), the corresponding values of $\varDelta x$ and $\varDelta y$ can be calculated with sufficient accuracy by multiplying Δs by $\cos \frac{1}{2}(\vartheta' + \vartheta'')$ for the former, and $\sin \frac{1}{2}(\vartheta' + \vartheta'')$ for the latter; or,

$$\Delta x = M \frac{k^2}{g} \log \frac{(i) - (\vartheta'')}{(i) - (\vartheta')} \cos \frac{1}{2}(\vartheta' + \vartheta'') = M \frac{k^2}{g} \Delta \xi \quad \text{(say)}$$

$$\Delta y = M \frac{k^2}{g} \log \frac{(i) - (\vartheta'')}{(i) - (\vartheta')} \sin \frac{1}{2}(\vartheta' + \vartheta'') = M \frac{k^2}{g} \Delta \zeta \quad \text{(say)}$$

For the entire range we evidently have

$$X = \Sigma \Delta x = M \frac{k^2}{g} \Sigma \Delta \xi = M \frac{k^2}{g} \xi$$

the summation extending from $\vartheta = \varphi$ to $\vartheta = \omega$, ω being the angle of fall.

To determine the value of ω we have, since the sum of the positive increments of g in the ascending branch is equal (numerically) to the sum of the negative increments in the descending branch,

$$\Sigma \Delta \zeta = 0.$$

Expression for the Time.—For the time of flight we have, when Δx is small,

$$\Delta t = \frac{\Delta x}{u}$$

in which u is the *mean* horizontal velocity corresponding to Δx; but, from (15), when $n = 2$,

$$u \doteq \frac{k}{\left\{ (i) - (\vartheta) \right\}^{\frac{1}{2}}}$$

whence

$$\Delta t = \frac{\Delta x \left\{ (i) - (\vartheta) \right\}^{\frac{1}{2}}}{k}$$

or, substituting for Δx its value given above,

$$\Delta t = \frac{Mk}{g} \Delta \xi \left\{ (i) - (\vartheta) \right\}^{\frac{1}{2}}$$

If we put

$$\Delta \theta = \Delta \xi \left\{ (i) - (\vartheta) \right\}^{\frac{1}{2}}$$

12

we may have

$$\log \varDelta\theta = \log \varDelta\xi + \tfrac{1}{2} \log [(i) - (\vartheta)]$$

The two values of $\log [(i) - (\vartheta)]$ corresponding to the extremities of the arc $\varDelta s$, are

$$\log [(i) - (\vartheta')], \text{ and } \log [(i) - (\vartheta'')]$$

the first of which is too small and the second too great; whence, taking their arithmetical mean,

$$\log \varDelta\theta = \log \varDelta\xi + \tfrac{1}{4} \log [(i) - (\vartheta')] + \tfrac{1}{4} \log [(i) - (\vartheta'')]$$

by means of which θ may be computed, and we then have

$$T = M \frac{k}{g} \theta$$

Tables.—General Otto, of the Prussian Artillery, has published extensive tables* of the values of (ϑ), ξ, ζ, and θ— the last three double entry tables with i and φ for the arguments—by means of which it is easy to solve many of the problems of high-angle fire.

Determination of k^2.—General Otto, in the work above cited, gives the following method for determining k^2: We have

$$X = M \frac{k^2}{g} \xi$$

and

$$T^2 = M' \frac{k^2}{g^2} \theta^2$$

whence

$$\frac{M}{g} \frac{X}{T^2} = \frac{\xi}{\theta^2}$$

an equation independent of k^2. Moreover ξ and θ^2 are both independent of X and T, being functions of the angle i and the angle of projection φ; and their ratio $\dfrac{\xi}{\theta^2}$ may be tabulated with these angles for arguments. General Otto has inserted such a table in his work calculated for angles of

---

* "Tafeln für den Bombenwurf." Translated into French by Rieffel with the title "Tables Balistiques Générales pour le tir élevé." Paris, 1844.

projection beginning at 30° and proceeding by intervals of 5° up to 75°.

Now, suppose a certain projectile is fired with a known angle of projection φ, and its horizontal range X, and time of flight T, are carefully measured. With this data we compute $\dfrac{\xi}{\theta^2}$ by means of the above equation; and entering Otto's Table III. with the argument φ, find in the proper column the computed value of $\dfrac{\xi}{\theta^2}$, and take out the corresponding value of i. Next, with φ and i as arguments, take from Table II. the value of ξ, from which k^2 can be computed by the following formula, derived from the expression for X given above :

$$ k^2 = \frac{g}{M} \frac{X}{\xi} $$

BASHFORTH'S METHOD.

For all values of n greater than unity the differential equations of motion take their simplest form when $n = 3$. For this reason Professor Bashforth assumes the cubic law of resistance throughout the whole extent of the trajectory, and employs variable coefficients to make the results conform to the actual resistance.

Making $n = 3$, equation (25) becomes

$$ dt = -\frac{k}{g} \frac{d \tan \vartheta}{\left\{ (i) - (\vartheta) \right\}^{\frac{1}{3}}} $$

in which

$$ (\vartheta) = \tan \vartheta + \tfrac{1}{3} \tan^3 \vartheta $$

From (14) we have, when $n = 3$ and $\vartheta = 0$,

$$ (i) = \frac{k^2}{v_0{}^3} $$

and this substituted in the above expression for dt gives, by a slight reduction,

$$dt = -\frac{v_0}{g} \left\{ 1 - \frac{v_0^3}{3k^3} (3 \tan \vartheta + \tan^3 \vartheta) \right\}^{\frac{1}{3}} d \tan \vartheta$$

Introducing Bashforth's coefficient K, making

$$\frac{K}{g} \frac{d^2}{w} \left(\frac{v_0}{1000}\right)^3 = \gamma$$

to correspond with his notation, and integrating between the limits (φ, ϑ) and $(0, t)$, we have

$$t = \frac{v_0}{g} \int_\theta^\phi \frac{d \tan \vartheta}{\left\{ 1 - \gamma (3 \tan \vartheta + \tan^3 \vartheta) \right\}^{\frac{1}{3}}} = \frac{v_0}{g} {}^\phi T_\gamma^\theta$$

Operating in the same way upon (26) and (27), we obtain

$$x = \frac{v_0^2}{g} \int_\theta^\phi \frac{d \tan \vartheta}{\left\{ 1 - \gamma (3 \tan \vartheta + \tan^3 \vartheta) \right\}^{\frac{2}{3}}} = \frac{v_0^2}{g} {}^\phi X_\gamma^\theta$$

and

$$y = \frac{v_0^2}{g} \int_\theta^\phi \frac{\tan \vartheta\, d \tan \vartheta}{\left\{ 1 - \gamma (3 \tan \vartheta + \tan^3 \vartheta) \right\}^{\frac{2}{3}}} = \frac{v_0^2}{g} {}^\phi Y_\gamma^\theta$$

Professor Bashforth has published extensive tables of the definite integrals ${}^\phi T_\gamma^\theta, {}^\phi X_\gamma^\theta$, and ${}^\phi Y_\gamma^\theta$ for values of ϑ extending from $+60°$ to $-60°$, and of γ from 0 to 100, calculated by quadratures; by means of which the principal elements of a trajectory may be accurately determined as follows:

As the coefficient of resistance K generally varies with the velocity, the trajectory must be divided into arcs of such limited extent that the value of K for each arc may be considered constant; and it should be so taken as to give, as nearly as possible, its mean value for the arc under consideration.

In the equation given on page 65, viz.:

$$\left(\frac{1000}{v_0}\right)^3 = \left(\frac{1000}{U}\right)^3 + \frac{K}{g} \frac{d^2}{w} \left\{ 3 \tan \varphi + \tan^3 \varphi \right\}$$

suppose U and φ to be the initial horizontal velocity and angle of projection respectively, and both known; and let ϑ, also known, be the inclination of the forward extremity of the first arc into which the trajectory is divided. Now, assuming a mean velocity for this arc, take out the corresponding value of K from the proper table and compute $\left(\dfrac{1000}{u_0}\right)^3$; then, in the same equation, changing φ to ϑ, U becomes the horizontal velocity at the forward extremity of the arc, which can also be computed.

Next compute γ by means of the equation given above, with which and the known values of φ and ϑ enter the tables and take out $^\phi T^\theta_\gamma$, $^\phi X^\theta_\gamma$, and $^\phi Y^\theta_\gamma$; lastly, multiplying the first by $\dfrac{v_0}{g}$, and each of the others by $\dfrac{v_0^2}{g}$, we have the time of describing the first arc of the trajectory and the co-ordinates of its forward extremity. By repeating the process with the second and following arcs into which the trajectory may be divided, the whole trajectory becomes known.

Professor Bashforth gives various other tables in his work, besides those we have mentioned, for facilitating the calculation of trajectories by his method, with examples of their application and full directions for their use.

Modification of Bashforth's Method for low Velocities.—When the initial velocity does not exceed 790 f. s. the law of resistance is that of the square of the velocity for the entire trajectory; and even when the initial velocity is as great as 1000 f. s. examples show that no material error results if we still retain the law of the square in our calculations; and this furnishes a very easy method for calculating trajectories for high angles of projection and for the initial velocities usually employed in high-angle fire, and which, it is believed, gives as accurate results as by any other method, and with less labor.

Making $n = 2$, equation (25) becomes

$$dt = -\frac{k}{g}\frac{d\tan\vartheta}{\left\{(i) - (\vartheta)\right\}^{\frac{1}{2}}}$$

in which

$$(\vartheta) = \frac{1}{2}\left\{\tan\vartheta\sec\vartheta + \log\tan\left(\frac{\pi}{4} + \frac{\vartheta}{2}\right)\right\}$$

We also have from (15), when $n=2$, and $\vartheta=0$,

$$(i)=\frac{k^2}{v_0^2}=\frac{1}{\gamma} \quad \text{(say)}$$

and this substituted in the above expression for dt gives

$$dt=-\frac{v_0}{g}\frac{d\tan\vartheta}{\left\{1-\gamma\left(\vartheta\right)\right\}^{\frac{1}{2}}}$$

whence

$$t=\frac{v_0}{g}\int_\theta^\phi\frac{d\tan\vartheta}{\left\{1-\gamma\left(\vartheta\right)\right\}^{\frac{1}{2}}}=\frac{v_0}{g}\,{}^\phi T_\gamma^\theta$$

In the same way we obtain from (26) and (27) the follow-ing expressions for x and y:

$$x=\frac{v_0^2}{g}\int_\theta^\phi\frac{d\tan\vartheta}{1-\gamma\left(\vartheta\right)}=\frac{v_0^2}{g}\,{}^\phi X_\gamma^\theta$$

and

$$y=\frac{v_0^2}{g}\int_\theta^\phi\frac{\tan\vartheta\,d\tan\vartheta}{1-\gamma\left(\vartheta\right)}=\frac{v_0^2}{g}\,{}^\phi Y_\gamma^\theta$$

It will be seen that this method depends upon tables of definite integrals which must be calculated by quadratures as in Bashforth's method, and with the same number of arguments; but the great advantage of these formulæ over Bashforth's is in the fact that γ is constant for a given trajectory, and, therefore, the labor of calculation is the same for all angles of projection.

To determine the value of k^2 for oblong projectiles of the standard type we have

$$k^2=\frac{gC}{2A}$$

Taking the value of A derived from the Bashforth experiments for velocities less than 790 f. s., and making $g=32.16$, we find

$$k^2=[5.4359033]\ C$$

For the Krupp projectiles we should have, taking Mayevski's value of A,

$$k^2=[5.5367564]\ C$$

The numbers between brackets are the logarithms of the factors by which C is to be multiplied.

For computing v_0 we have from (32), when $\vartheta = 0$,

$$I(v_0) = \frac{2}{C}(\varphi) + I(U) \qquad (56)$$

in which φ may be the inclination at any point in either branch, and U the corresponding horizontal velocity. The values of (φ) are given in Table III.

To show the practical working of this method, we will take the example from Bashforth already given (see page 67). The data are: $V = 751$ f. s.; $\varphi = 30°$; $d = 6.27$ inches, and $w = 70$ lbs.; whence $U = 650.385$ f. s., and $C = \dfrac{70}{(6.27)^2} = 1.78059$. Determine the range, time of flight, angle of fall, and terminal velocity.

First compute v_0. We have from Table III.

$$(30°) = 0.60799$$

whence, from (56),

$$I(v_0) =$$
$$\frac{2 \times 0.60799}{1.78059} + I(650.385) = 0.68291 + 0.93354 = 1.61645$$

therefore, from Table I.,

$$v_0 = 525.91 \text{ f. s.}$$

Computation of γ:

$$
\begin{aligned}
\log C &= 0.2505630 \\
\text{constant } \log &= 5.4359033 \\
\log k^2 &= 5.6864663 \\
\log v_0^2 &= 5.4418228 \\
\log \gamma &= 9.7553565 \\
\gamma &= 0.56932
\end{aligned}
$$

As general tables of the definite integrals $^\phi T_\gamma^\theta$, $^\phi X_\gamma^\theta$, and $^\phi Y_\gamma^\theta$ have not yet been prepared, the following table has been calculated for this particular example, merely to illustrate the method:

$$\gamma = 0.56932$$

θ	T	X	Y
30°	0.63676	0.70486	0.21775
24	.47838	.51493	.12039
18	.34169	.35965	.06045
12	.21944	.22662	.02460
+ 6	+ .10673	+ .10838	+ .00575
0	.00000	.00000	.00000
− 6	− .10358	− .10208	+ .00531
12	.20647	.20061	.02091
18	.31104	.29793	.04701
24	.41977	.39620	.08479
30	.53551	.49759	.13656
36	.66179	.60449	.20615
37	.68417	.62303	.21987

The value of $^{30°}Y^{o}_{\gamma}$ by the above table is 0.21775, and as this must be equal to $^{o}Y^{\omega}_{\gamma}$ we see at a glance that ω lies between $-36°$ and $-37°$; and by interpolation we get $\omega = -36° 51'$; and therefore $^{o}X^{\omega}_{\gamma} = 0.62025$ and $^{o}T^{\omega}_{\gamma} 0.68081$. Adding to these the numbers corresponding to the argument 30°, we get $^{\phi}X^{\omega}_{\gamma} = 1.32511$, and $^{\phi}T^{\theta}_{\gamma} = 1.31757$. Lastly, multiplying the first of these by $\dfrac{v_{o}^{2}}{g}$, and the second by $\dfrac{v_{o}}{g}$, we obtain

$$X = 11396 \text{ ft.}$$

and

$$T = 21''.546$$

which agree with Bashforth's calculations.

The terminal velocity is found from (32), viz.:

$$I(u_{\omega}) = \frac{2}{C}(\omega) + I(v_{o})$$

and

$$v_{\omega} = u_{\omega} \sec \omega$$

We find

$$u_{\omega} = 434.7 \text{ f. s.}$$

and

$$v_{\omega} = 543.2 \text{ f. s.}$$

It will be seen that the inverse problem, namely, Given

the terminal velocity and angle of fall, to determine the initial velocity, angle of projection, range, and time, can be solved by this method with the same ease and accuracy as the direct problem. We should first compute the summit velocity by the equation

$$I (v_0) = I (u_\omega) - \frac{2}{C} (\omega) \tag{57}$$

and then all the other elements would be determined, as already explained.

In calculating trajectories by this method with the help of tables of the definite integrals $^\phi T_\gamma^\theta$, etc., it will generally be necessary, as in Bashforth's method, to interpolate with reference to γ as well as ϑ, and for this purpose the integrals must be tabulated for different values of γ proceeding by constant differences, and including the highest and lowest values of γ likely to be needed in practice, which are, approximately, I and 0.2.

13

CHAPTER VII.

TRAJECTORIES CONTINUED—DIRECT FIRE.

Niven's Method.—If a is some mean value of sec ϑ between the limits of integration; that is, if we make
$$a = \sec \bar{\vartheta} \quad \text{(say)}$$
then equations (17) to (20) may be written as follows:

$$dt = -\frac{C}{A}\frac{d\,(a\,u)}{(a\,u)^n}$$

$$dx = -\frac{C}{A}\cos\bar{\vartheta}\,\frac{d\,(a\,u)}{(a\,u)^{n-1}} \qquad (58)$$

$$dy = -\frac{C}{A}\sin\bar{\vartheta}\,\frac{d\,(a\,u)}{(a\,u)^{n-1}}$$

$$ds = -\frac{C}{A}\frac{d\,(a\,u)}{(a\,u)^{n-1}}$$

Making $a\,u = u'$, and integrating so that t, x, y, and s shall each be zero when $u' = U'$, we have

$$t = \frac{C}{(n-1)\,A}\left\{\frac{1}{u'^{n-1}} - \frac{1}{U'^{n-1}}\right\}$$

$$x = \frac{C}{(n-2)\,A}\cos\bar{\vartheta}\left\{\frac{1}{u'^{n-2}} - \frac{1}{U'^{n-2}}\right\}$$

$$y = \frac{C}{(n-2)\,A}\sin\bar{\vartheta}\left\{\frac{1}{u'^{n-2}} - \frac{1}{U'^{n-2}}\right\}$$

$$s = \frac{C}{(n-2)\,A}\left\{\frac{1}{u'^{n-2}} - \frac{1}{U'^{n-2}}\right\}$$

Comparing these equations with those deduced in Chapter IV. for rectilinear motion, it will be evident that we have as follows:

$$t = C\,[T\,(u') - T\,(U')] \qquad (59)$$
$$x = C\cos\bar{\vartheta}\,[S\,(u') - S\,(U')] \qquad (60)$$
$$y = C\sin\bar{\vartheta}\,[S\,(u') - S\,(U')] = x\tan\bar{\vartheta} \qquad (61)$$
$$s = C\,[S\,(u') - S\,(U')] \qquad (62)$$

The first three of these equations (or their equivalents) were first published by Mr. Niven in 1877, and in connection with equation (38), viz.:

$$D = C \cos \bar{\vartheta} \, [D\,(u') - D\,(U')] \qquad (63)$$

constitute what is known as "Niven's Method."

If we use the I-function instead of the D-function, equation (63) becomes

$$D = \frac{90}{\pi} C \cos \bar{\vartheta} [I\,(u') - I\,(U')] \qquad (64)$$

or, better still, for direct fire (see Chapter V.),

$$D = \frac{90}{\pi} C \sec \varphi \, [I\,(u \sec \varphi) - I\,(V)] \qquad (65)$$

in which

$$\log \frac{90}{\pi} = 1.4570926 \cdot$$

The values of $\bar{\vartheta}$ adopted by Mr. Niven are as follows:
For the D-integral

$$\tan \bar{\vartheta}_1 = \frac{\tan \varphi + \tan \vartheta}{2}$$

For the X-, Y-, and T-integrals

$$\bar{\vartheta} = \bar{\vartheta}_1 + \frac{U - u}{U + u} \frac{\varphi - \vartheta}{3}$$

for the ascending branch, and

$$\bar{\vartheta} = \bar{\vartheta}_1 - \frac{U - u}{U + u} \frac{\vartheta - \varphi}{3}$$

for the descending branch of the trajectory. For the method of deducing these expressions for $\bar{\vartheta}$, see a paper by Professor J. M. Rice, U. S. Navy, in the eighth volume of "Proceedings Naval Institute," page 191.

We will now apply these formulæ to the solution of a problem of direct fire; and, as we wish to compare the results obtained with those to be deduced from other methods we will use Table I. of this work instead of Niven's tables, and we will also perform the calculations with more accuracy than is generally necessary in practice.

Example of Niven's Method.—A 12-inch service projectile is fired at an angle of departure of 10°, and an initial velocity of 1886 f. s. Find v, x, y, and t (a) when $\vartheta = 0$, and (b) when $\vartheta = -13°$.

Here $d = 12$ in., $w = 800$ lbs., $C = \dfrac{800}{144}$, $\varphi = 10°$, $V = 1886$ f. s., $U = 1886 \cos 10° = 1857.33$.

(a) $\vartheta = 0$ $\therefore D = 10°$. We have first

$$\tan \overline{\vartheta}_1 = \tfrac{1}{2} \tan 10° = 0.0831635$$
$$\therefore \overline{\vartheta}_1 = 5° 2' 18'', \text{ and } U' = U \sec \overline{\vartheta}_1 = 1864.56$$

Next compute u' by means of the equation

$$I(u') = \frac{\pi}{90} \frac{D}{C} \sec \overline{\vartheta}_1 + I(U')$$

or

$$I(u') = 0.06308 + 0.03624 = 0.09932$$
$$\therefore u' = 1328.96 = u_o \sec \overline{\vartheta}_1$$
$$\therefore u_o = 1323.72$$

Next compute the value of $\overline{\vartheta}$ to be used with the X-, Y-, and T-integrals. We have

$$\overline{\vartheta} = 5° 2' 18'' + \frac{1857.33 - 1323.72}{1857.33 + 1323.72} \times \frac{10}{3} = 5° 35' 51''$$

The new values of U' and u' are, therefore,

$$U' = 1866.25, \text{ and } u' = 1330.06$$

From Table I. we find

$$S(U') = 2855.3 \qquad S(u') = 5239.2$$
$$T(U') = 1.258 \qquad T(u') = 2.778$$

$$\therefore t_o = \frac{800}{144} \left\{ 2.778 - 1.258 \right\} = 8''.444$$

$$x_o = \frac{800}{144} \cos \overline{\vartheta} \left\{ 5239.2 - 2855.3 \right\} = 13180.7 \text{ ft.}$$

$$y_o = x \tan \overline{\vartheta} = 1291.8 \text{ ft.}$$

(b) $\vartheta = -13°$. It will be necessary in this case to take a new origin at the summit of the trajectory, as there is no provision made in this method for calculating an arc of a trajectory lying partly in the ascending and partly in the descending branches. Indeed, since the differential expression for y contains sin ϑ as a factor, which becomes zero at the summit and changes its sign in the descending branch, equation (61) does not hold true, unless the limits of integration (φ and ϑ) are both positive or both negative.

We have, then, for this arc of the trajectory the following data:

$$V = U = 1323.72, \varphi = 0°, \vartheta = -13°, \text{ and } D = 13°$$

$$\tan \bar{\vartheta}_1 = -\tfrac{1}{3} \tan 13° = -0.1154341 \quad \therefore \bar{\vartheta}_1 = -6° 35' 5''$$

$$U' = 1332.51 \quad I(U') = 0.09860$$

$$I(u') = 0.08222 + 0.09860 = 0.18082$$

$$\therefore u' = 1064.39 = v_\theta \cos \vartheta \sec \bar{\vartheta}_1$$

$$\therefore v_\theta = 1085.18, \text{ and } u_\theta = 1057.37$$

$$\bar{\vartheta} = -6° 35' 5'' - \frac{1323.72 - 1057.37}{1323.72 + 1057.37} \times \frac{-13}{3} = -6° 6' 0''$$

The new values of U' and u' are, therefore,

$$U' = 1331.26, \text{ and } u' = 1063.39.$$

From Table I. we get

$$S(U') = 5232.9 \qquad S(u') = 7011.7$$
$$T(U') = 2.773 \qquad T(u') = 4.282$$

$$\therefore t = \frac{800}{144} \left\{ 4.282 - 2.773 \right\} = 8''.383$$

$$x = \frac{800}{144} \cos \bar{\vartheta} \left\{ 7011.7 - 5232.9 \right\} = 9826.3 \text{ ft.}$$

$$y = x \tan \bar{\vartheta} = -1050.1$$

The co-ordinates of the point of the trajectory whose inclination is $-13°$, taking the origin at the point of projection, are therefore

$$X = 13180.7 + 9826.3 = 23007.0 \text{ ft.}$$
$$Y = 1291.8 - 1050.1 = 241.7 \text{ ft.}$$

And the time,

$$T = 8.444 + 8.383 = 16''.827$$

For comparison we have computed the same elements

directly from equations (16), (25), (26), and (27), dividing the whole arc into three parts, with the points of division corresponding to velocities of 1330 f. s. and 1120 f. s. respectively. The integrals for each arc were computed by quadratures, and the following are the final results:

$$v_\theta = 1081.55 \text{ f. s.}; \quad X = 23025.7 \text{ ft.}; \quad Y = 243.14 \text{ ft., and}$$
$$T = 16''.843.$$

The agreement between these two sets of values is remarkably close, and shows that for the purpose of computing co-ordinates of different points of a trajectory, Niven's method is all that could be desired so far as accuracy is concerned. For high angles of projection the trajectory should be divided into arcs not exceeding 10° or 15° each, and always with one point of division at the summit.

Example 2.—Given $d = 12$ in., $w = 800$ lbs., $V = 1886$ f. s., and $\varphi = 30°$. Compute the time and co-ordinates when $\vartheta = 24°$.

Answer:

BY NIVEN'S METHOD.	BY QUADRATURES.
$\overline{\vartheta_1} = 27° \ 4' \ 29''$	
$\overline{\vartheta} = 27° \ 19' \ 4''$	
$x_\theta = 8482.0$ ft.	8481.4 ft.
$y_\theta = 4381.2$ ft.	4381.9 ft.
$t_\theta = 5''.889$	5''.888
$v_\theta = 1400.58$ f. s.	1400.4 f. s.

In the same manner, by successive steps, can the whole trajectory be computed. In practice it is never necessary to divide a trajectory into arcs of less than 10°.

Sladen's Method for Low-Angle Firing.*—When the angle of projection is small, say not exceeding 3°, the time corresponding to a given range can be computed with great accuracy by means of (29) and (30). We should first find v by means of the equation

$$S(v) = \frac{X}{C} + S(V)$$

---

* "Principles of Gunnery," by Major J. Sladen, R.A., London, 1879, Chapter VI.

and then with this value of v compute T by means of (29). In the same manner we could find the value of t for a given value of x, less than X; and these values of T and t substituted in (46), viz.,

$$y = \frac{g\,t}{2}(T - t)$$

would give the value of y corresponding to x; since, under the conditions supposed, the vertical component of the velocity would be so small as to produce no appreciable resistance to the projectile in that direction.

Example 1.—Required the following co-ordinates of the trajectory described by a 500-grain bullet fired from a Springfield rifle, for a range of 600 ft., viz.: when $x = 150$ ft., 300 ft., and 450 ft. respectively; $\delta = 524.29$, $\delta_1 = 534.22$.

Here $d = 0.45$ in., $w = 500$ grains $= \frac{1}{14}$ lb., $V = 1280$ f. s., and $X = 600$ ft. We first find $S(V) = 5509.70$; $T(V) = 2.985$; and

$$C = \frac{\frac{1}{14} \times 534.22}{(0.45)^2 \times 524.29} = 0.35942$$

The principal steps of the remaining calculations are given in the following table:

X (ft.)	$\frac{X}{C}$	$S(v)$	v (f. s.)	t	y (inches.)	y' (inches.)	y_0 (inches.)
150	417.34	5727.04	1209.72	$0''.12055$	9.365	9.406	7.950
300	834.69	6344.39	1146.76	$0''.24814$	13.167	12.987	10.600
450	1252.03	6761.73	1091.31	$0''.38235$	10.386	9.956	7.950
600	1669.38	7179.08	1046.55	$0''.52313$ (T)	0.000	0.000	0.000

The sixth column gives the computed values of y, and the seventh the mean of five trajectories measured with great care at Creedmoor by Mr. H. G. Sinclair, in charge of the "Forest and Stream Trajectory Test." The last column gives the corresponding values of y *in vacuo*, computed by (45).

SIACCI'S METHOD FOR DIRECT FIRE.

Expression for y.—We have from (35), since $\tan \vartheta = \dfrac{dy}{dx}$

$$\frac{dy}{dx} = \tan \varphi - \frac{a\,C}{2}\left\{ I(u') - I(U') \right\}$$

or

$$\frac{2}{a\,C}\left\{ \frac{dy}{dx} - \tan \varphi \right\} - I(U') = -I(u')$$

We also have from (58)

$$\frac{a}{C}\,dx = -\frac{du'}{A\,u'^{n-1}}$$

whence multiplying the last two equations together, member by member,

$$\frac{2}{C^2}\left\{ dy - \tan \varphi\,dx \right\} - \frac{a}{C}I(U')\,dx = \frac{I(u')\,du'}{A\,u'^{n-1}}.$$

Integrating and making x and y both zero at the origin, where $u' = U'$, we have

$$\frac{2}{C^2}\left\{ y - x\tan \varphi \right\} - \frac{a}{C}I(U')\,x = \frac{1}{A}\int_{U'}^{u'}\frac{I(u')\,du'}{u'^{n-1}}$$

Making for convenience

$$-A(u') = \frac{1}{A}\int\frac{I(u')\,du'}{u'^{n-1}}$$

(in which the A's must not be confounded) the above equation becomes

$$\frac{2}{C^2}\left\{ y - x\tan \varphi \right\} - \frac{a}{C}I(U')\,x = -\left\{ A(u') - A(U') \right\}$$

From (60) we have

$$\frac{a}{C}x = S(u') - S(U')$$

whence, by division,

$$\frac{2}{a\,C}\left\{ \frac{y}{x} - \tan \varphi \right\} - I(U') = -\frac{A(u') - A(U')}{S(u') - S(U')}$$

or

$$\frac{y}{x} = \tan \varphi - \frac{a\,C}{2}\left\{ \frac{A(u') - A(U')}{S(u') - S(U')} - I(U') \right\} \qquad (66)$$

Calculation of the A-Function.—We have (Chapter V.)

$$I(u') = \frac{2g}{n A u'^n} + Q$$

and therefore

$$A(u') = -\frac{2g}{n A^2}\int \frac{du'}{u'^{2n-1}} - \frac{Q}{A}\int \frac{du'}{u'^{n-1}} + Q'$$

$$= \frac{g}{n(n-1)A^2 u'^{2(n-1)}} + \frac{Q}{(n-2)A u'^{n-2}} + Q'$$

which becomes, when $n = 2$,

$$A(u') = \frac{g}{2 A^2 u'^2} - \frac{Q}{A}\log u' + Q'$$

The constants Q, corresponding to the five different expressions for the resistance, are given in Chapter V., and the values of Q' are to be determined as explained in Chapter IV. Making the necessary substitutions, and using $A(v)$ as the general functional symbol, we have for standard oblong projectiles the following expressions for calculating the A-functions:

2800 f. s. $> v >$ 1330 f. s.:

$$A(v) = [8.9012292]\frac{1}{v^3} + [2.6701589]\log v - 1714.55$$

1330 f. s. $> v >$ 1120 f. s.:

$$A(v) = [14.6562945]\frac{1}{v^4} + [5.1480576]\frac{1}{v} - 53.13$$

1120 f. s. $> v >$ 990 f. s.:

$$A(v) = [32.2571789]\frac{1}{v^{10}} + [14.4412953]\frac{1}{v^4} + 126.68$$

990 f. s. $> v >$ 790 f. s.:

$$A(v) = [14.9781903]\frac{1}{v^4} - [5.9124902]\frac{1}{v} + 449.89$$

790 f. s. $> v >$ 100 f. s.:

$$A(v) = [9.6655206]\frac{1}{v^3} + [4.1438598]\log v - 45916.40$$

The values of $A(v)$ calculated by the above formulæ are given in Table I.

14

Equation (66), together with (35), (59), and (60), are the fundamental equations of " Siacci's method." This method, by Major F. Siacci, of the Italian Artillery, was published in the *Revue d'Artillerie* for October, 1880. A translation of this paper by Lieutenant O. B. Mitcham, Ordnance Department, U. S. A., was printed in the report of the Chief of Ordnance for 1881. Lieutenant Mitcham added to his translation a ballistic table adapted to English units, and based upon the coefficients of resistance deduced by General Mayevski from the Russian and English experiments noticed in Chapter II. In this table he gives for the first time the values of $T(v)$.

We will, for convenience, collect these equations together and renumber them:

They are:

$$\tan \varphi - \tan \vartheta = \frac{aC}{2} \left\{ I(u') - I(U') \right\} \tag{67}$$

$$x = \frac{C}{a} \left\{ S(u') - S(U') \right\} \tag{68}$$

$$\frac{y}{x} = \tan \varphi - \frac{aC}{2} \left\{ \frac{A(u') - A(U')}{S(u') - S(U')} - I(U') \right\} \tag{69}$$

$$t = C[T(u') - T(U')] \tag{70}$$

$$u' = a v \cos \vartheta \tag{71}$$

As the origin of co-ordinates is at the point of departure, y is zero at the origin and also at the point in the descending branch where the trajectory pierces the horizontal plane passing through the muzzle of the gun. Calling the velocity at this point v_ω, we shall have, making $-\vartheta = \omega$,

$$u'_\omega = a v_\omega \cos \omega \tag{72}$$

From (69) we have

$$\tan \varphi = \frac{aC}{2} \left\{ \frac{A(u'_\omega) - A(U')}{S(u'_\omega) - S(U')} - I(U') \right\} \tag{73}$$

and from (67)

$$\tan \varphi = \frac{aC}{2} \left\{ I(u'_\omega) - I(U') \right\} - \tan \omega \tag{74}$$

Eliminating $\tan \varphi$ from these last two equations gives

$$\tan \omega = \frac{aC}{2} \left\{ I(u'_\omega) - \frac{A(u'_\omega) - A(U')}{S(u'_\omega) - S(U')} \right\} \quad (75)$$

From (68) and (70) we have

and

$$X = \frac{C}{a} \left\{ S(u'_\omega) - S(U') \right\} \quad (76)$$

$$T = C[T(u'_\omega) - T(U')] \quad (77)$$

By means of equations (67) to (77) all problems of exterior ballistics in the plane of fire may be solved. If we wish to compute the co-ordinates of the extremities of any arc of a trajectory having the inclinations φ and ϑ, we should make use of equations (67) to (71). If the object is to determine the elements of a complete trajectory lying above the horizontal plane passing through the muzzle of the gun, at one operation, we should employ equations (72) to (77). We will give an example of each, using Didion's value of a.

Example 1.—Given $V = 1886$ f. s.; $d = 12$ in.; $w = 800$ lbs., $\varphi = 10°$, and $\vartheta = -13°$; to find v_θ, x_θ, y_θ, and t_θ. (See example 1, Niven's method.)

We have first

$$a = \frac{(10°) + (13°)}{\tan 10° + \tan 13°} = 1.007231$$

Next

$$U' = 1886\, a \cos 10° = 1870.78$$

From Table I.,

$S(U') = 2838.3$; $A(U') = 44.06$; $I(U') = 0.03581$; $T(U') = 1.250$

From (67) we have

$$I(u'_\theta) = \frac{2}{aC} \left\{ \tan 10° + \tan 13° \right\} + I(U')$$

$$= 0.14554 + 0.03581 = 0.18135$$

$\therefore u' = 1063.42$; $S(u') = 7011.4$; $A(u') = 440.44$; $T(u') = 4.282$.

These values substituted in (68), (69), and (70) give

$$x_\theta = 23017 \text{ ft.}$$
$$y_\theta = 248.06 \text{ ft.}$$
$$t_\theta = 16''.844$$

From (71) we have

$$v_\theta = \frac{u'}{a \cos \vartheta} = 1083.6 \text{ f. s.}$$

These results are quite as accurate as those deduced by Niven's method by two steps.

Example 2.—Required the horizontal range, time of flight, and striking velocity, with the data of Example 1.

In computing a we will assume an angle of fall of $-14° 30'$, which gives

$$a = 1.008645$$
$$\therefore U' = 1873.40$$

$S(U')=2828.5$; $A(U')=43.71$; $I(U')=0.03563$; $T(U')=1.243$.

From (73) we have

$$\frac{A(u'_\omega) - 43.71}{S(u'_\omega) - 2828.5} = \frac{2}{a C} \tan \varphi + I(U') = 0.09856$$

from which to calculate u'_ω. As the relation between the S-function and A-function does not admit of a direct solution of this equation, it will be necessary to determine the value of u'_ω by successive approximations; and for this purpose the rule of " Double Position " is well adapted. This rule is deduced as follows : Let u_1 and u_2 be two near values of u (or the quantity to be determined), one greater and the other less ; and e_1 and e_2 the errors respectively, when u_1 and u_2 are substituted for u in the equation to be solved. Then, upon the hypothesis that the errors in the results are proportional to the errors in the assumed data, we have

$$e_1 : e_2 :: u - u_1 : u - u_2$$

whence, by division,

$$e_1 - e_2 : e_1 :: u_2 - u_1 : u - u_1$$

or

$$e_1 - e_2 : e_2 :: u_2 - u_1 : u - u_2$$

from which is derived the following rule: As the difference of the errors is to the difference of the assumed numbers, so is the lesser of the two errors (numerically) to the correction to be applied to the corresponding assumed number.

If u_1 and u_2 are selected with judgment, the resulting value of u will generally be sufficiently correct by a single application of the rule, or, at most, by two trials.

In our example assume $u_1 = 1050$, for a first trial; whence $S(1050) = 7143.7$, and $A(1050) = 464.94$; and these in the above equation give

$$\frac{464.94 - 43.71}{7143.7 - 2828.5} = 0.09762$$

If we had taken for u_1 the correct value of u'_ω, the second member would have been 0.09856, and hence $e_1 = -0.00094$. Whenever e_1 is negative the assumed value of u'_ω is too great; we will, therefore, next suppose $u_2 = 1040$, and proceeding in the same way we find $e_2 = +0.00128$. The correct value of u'_ω is, then, between 1050 ft. and 1040 ft. Applying the rule, we have the following proportion:

$$222 : 10 :: 94 : 4.23$$

consequently $u'_\omega = 1050 - 4.23 = 1045.77$ f. s.; and this satisfies the above equation.

We next find

$$S(u'_\omega) = 7187.1; \; A(u'_\omega) = 473.20; \; I(u'_\omega) = 0.19154; \; T(u'_\omega) = 4.448$$

We now have from (75)

$$\tan \omega = \frac{aC}{2} \left\{ 0.19154 - 0.09856 \right\} = 0.26051$$

$$\therefore \omega = 14° 36'. \quad \text{(By Table III.)}$$

From (76) and (77)

$$X = \frac{C}{a} \left\{ 7187.1 - 2828.5 \right\} = 24007 \text{ ft.}$$

$$T = C [4.448 - 1.243] = 17''.806$$

From (72)

$$v_\omega = \frac{u'_\omega}{a \cos \omega} = 1071.4 \text{ f. s.}$$

Various other problems may be solved by a suitable combination of equations (67) to (71). Indeed, if a velocity, either initial or terminal, and one other element be given, all the other elements may be computed, though in certain cases this can only be accomplished by successive approximations. Most of these problems, *for direct fire*, will be solved further on.

Application of Siacci's Equations to Mortar-Firing.—For low velocities, such as are used in mortar-firing, we may take for a in all cases the following value:

$$a = \frac{(\varphi)}{\tan \varphi}$$

This simplifies the calculations, and gives results sufficiently accurate for most practical purposes, as the following examples will show:

Example 1.—Given $V = 751$ f. s.; $\varphi = 30°$; and log $C = 0.25056$. Required X, T, ω, and v_ω. (See Example 1, Chapter VI.)

We have, Table III., $(\varphi) = 0.60799$.

$$\log (\varphi) = 9.78390$$
$$\log \tan \varphi = 9.76144$$
$$\log a = 0.02246$$
$$\log V = 2.87564$$
$$\log \cos \varphi = 9.93753$$
$$\log U' = 2.83563 \qquad U' = 684.90$$

$S(U') = 13681.1$; $A(U') = 3444.43$; $I(U') = 0.80679$; $T(U') = 12.274$.

$$\log 2 = 0.30103 \qquad \text{[Equation (73)]}$$
$$c.\ \log a = 9.97754$$
$$c.\ \log C = 9.74944 \qquad \text{(add log tan } \varphi)$$
$$\log 0.61581 = 9.78945$$
$$I\ (U') = 0.80679$$
$$\overline{\qquad 1.42260}$$

$$\therefore \frac{A\ (u'_\omega) - 3444.43}{S\ (u'_\omega) - 13681.1} = 1.42260$$

By double position we find from this equation

$$u'_\omega = 459.78$$

$$\therefore S(u'_\omega) = 20443.1 \; ; \; I(u'_\omega) = 2.22481 \; ; \; T(u'_\omega) = 24.404$$

$$X = \frac{C}{a} \left\{ 20443.1 - 13681.1 \right\} = 11434 \text{ ft.}$$

$$T = C[24.404 - 12.274] = 21''.60$$

$$\tan \omega = \frac{aC}{2} \left\{ 2.22481 - 1.42260 \right\} \qquad \text{[Eq. (75)]}$$

$$\therefore \omega = 36° 57'$$

$$v_\omega = \frac{u'_\omega}{a \cos \omega} = 546.3 \text{ f. s.} \qquad \text{[Eq. (72)]}$$

Example 2.—Given $V = 977.71$ f. s., $\varphi = 35° 21'$, and log $C = 0.38722$. Required X, T, ω, and v_ω. (See Example 3, Chapter VI.)
Answer:

$$X = 19328 \text{ ft.}$$
$$T = 31''.63$$
$$u'_\omega = 517.63$$
$$\omega = 44° 44'$$
$$v_\omega = 675.65 \text{ f. s.}$$

Example 3.—Given $V = 609.63$ f. s.; $\varphi = 45°$, and log $C = 0.56809$; required X, T, ω, and v_ω. (See Example 5, Chapter VI.)
Answer:

$$X = 11984 \text{ ft.}$$
$$T = 28''.30$$
$$u'_\omega = 436.52$$
$$\omega = 49° 10'$$
$$v_\omega = 581.64$$

Siacci's Equations for Direct Fire.—As already stated, a is some mean value of the secants of the inclinations of the extremities of the arc of the trajectory over which we integrate; and consequently if we take the whole trajectory lying above the level of the gun, a will be greater than 1 and less than sec ω. To illustrate, suppose we have for our data a given projectile fired with a certain known initial velocity and angle of projection, and we wish to calculate the angle of fall, terminal velocity, range, and time of flight. If we calculate these elements by means of (75), (72), (76), and (77), making $a = 1$, they will be too great; while if a is made equal to sec ω, or even sec φ, they will be too small; and the correct value of each element would be found by giving to a some value intermediate to the two. Moreover, the value of a which would give the exact range would not give the exact time of flight or terminal velocity. These principles are further illustrated by the following numerical results, calculated from the data, $V = 1404$ f. s.; $\varphi = 10°$; $w = 183$ lbs., and $d = 8$ in.:

$a = 1$	$a = $ sec φ
$X = 13752$ ft.	$X = 13622$ ft.
$v_\omega = 892.2$ f. s.	$v_\omega = 881.4$ f. s.
$\omega = -13° 17'$	$\omega = -13° 23'$
$T = 13''.04$	$T = 12''.55$

As the true values of these elements lie between those we have computed, it will be seen that either set of values is correct enough for most purposes. It is, therefore, apparent that in direct fire we may give to a that value which shall reduce the above equations to their simplest forms, provided it lies between the limits $a = 1$ and $a = $ sec φ.

As we have already seen (Chapter V.), Major Siacci gives to a the value

$$a = (\sec \varphi)^{\frac{n-2}{n-1}}$$

by means of which equation (37) was obtained, viz.:

$$\tan \vartheta = \tan \varphi - \frac{C}{\cos^3 \varphi} \left\{ I(u') - I(V) \right\} \tag{78}$$

in which

$$u' = v \frac{\cos \vartheta}{\cos \varphi}$$

Making the same substitution in (68), (69), and (70), they become respectively

$$x = C [S (u') - S (V)] \tag{79}$$

$$\frac{y}{x} = \tan \varphi - \frac{C}{2 \cos^2 \varphi} \left\{ \frac{A (u') - A (V)}{S (u') - S (V)} - I (V) \right\} \tag{80}$$

$$t = \frac{C}{\cos \varphi} \left\{ T (u') - T (V) \right\} \tag{81}$$

When φ and ϑ are so small that the ratio of their cosines does not differ much from unity, we may put

$$u' = v$$

and the above equations become

$$\tan \vartheta = \tan \varphi - \frac{C}{2 \cos^2 \varphi} \left\{ I (v) - I (V) \right\} \tag{82}$$

$$x = C [S (v) - S (V)] \tag{83}$$

$$\frac{y}{x} = \tan \varphi - \frac{C}{2 \cos^2 \varphi} \left\{ \frac{A (v) - A (V)}{S (v) - S (V)} - I (V) \right\} \tag{84}$$

$$t = \frac{C}{\cos \varphi} \left\{ T (v) - T (V) \right\} \tag{85}$$

We shall retain this form of the ballistic equations in what follows, though when very accurate results are desired we must use u' instead of v.

When $y = 0$, we have from (84)

$$\sin 2\varphi = C \left\{ \frac{A (v) - A (V)}{S (v) - S (V)} - I (V) \right\}. \tag{86}$$

Substituting for $\tan \varphi$ in (84) its value from (82), and reducing, we have, when $y = 0$,

$$2 \cos^2 \varphi \tan \omega = C \left\{ I (v) - \frac{A (v) - A (V)}{S (v) - S (V)} \right\}$$

For small angles of projection we may put

$$2 \cos^2 \varphi \tan \omega = 2 \sin \omega \cos \omega \frac{\cos^2 \varphi}{\cos^2 \omega} = \sin 2\omega$$

and, therefore,

$$\sin 2\omega = C \left\{ I (v) - \frac{A (v) - A (V)}{S (v) - S (V)} \right\} \tag{87}$$

For the larger angles of projection employed in direct

15

fire, if accurate results are desired, we must determine ω by the equation

$$\tan \omega = \tan \varphi - \frac{C}{2 \cos^2 \varphi} \left\{ I(v) - I(V) \right\}$$

using u' instead of v, as already explained.

Practical Applications.—We will now apply Siacci's equations to the solution of some of the most important problems of direct fire.

PROBLEM I.—*Given the initial velocity and angle of projection, to determine the range, time of flight, angle of fall, and terminal velocity.*

We have [equation (86)]

$$\frac{A(v) - A(V)}{S(v) - S(V)} = \frac{\sin 2\varphi}{C} + I(V)$$

from which to calculate v by " Double Position," as already explained. Having found v, the remaining elements are computed by the equations

$$X = C \left[S(v) - S(V) \right]$$

$$T = \frac{C}{\cos \varphi} \left\{ T(v) - T(V) \right\}$$

$$\sin 2\omega = C \left\{ I(v) - \frac{A(v) - A(V)}{S(v) - S(V)} \right\}$$

For curved fire we may proceed as follows: We have, from the origin to the summit,

$$t_0 = \frac{C}{\cos \varphi} \left\{ T(v_0) - T(V) \right\}$$

Now, if we assume that the time from the point of projection to the summit is one-half the time of flight, we shall have, from the above expressions for T and t_0,

$$T(v) = 2\,T(v_0) - T(V)$$

which gives v by means of the T-functions, v_0 being computed by the equation

$$I(v_0) = \frac{\sin 2\varphi}{C} + I(V)$$

derived from (82).

Example 1.—The 8-inch rifle (converted) fires an ogival-headed shot weighing 183 lbs. If the angle of projection is 10°, and the initial velocity 1404 f. s., find the range, time of flight, angle of fall, and terminal velocity.

We have $V = 1404$ f. s.; $\varphi = 10°$; $w = 183$ lbs.; $d = 8$ inches, whence log $C = 0.45627$: to find X, T, ω, and v.

From Table I. we find

$$S(V) = 4878.6 - 0.8 \times 25.1 = 4858.5$$
$$A(V) = 163.96 - 0.8 \times 2.16 = 162.23$$
$$I(V) = 0.08661 - 0.8 \times 0.00082 = 0.08599$$
$$T(V) = 2.514 - 0.8 \times 0.018 = 2.500.$$

Next compute v:

$$\log \sin 2\varphi = 9.53405$$
$$\log C = 0.45627$$
$$\overline{\log 0.11961 = 9.07778}$$
$$I(V) = 0.08599$$
$$\overline{0.20560}$$
$$\therefore \frac{A(v) - A(V)}{S(v) - S(V)} = 0.20560$$

The value of v satisfying this equation is found to be $v = 873.8$ ft., whence

$$S(v) = 9641.8 \qquad A(v'_\omega) = 1145.65$$
$$I(v) = 0.36668 \qquad T(v'_\omega) = 7.030$$

X, T, ω, and v are now computed as follows:

$$\log C = 0.45627$$
$$\log [S(v) - S(V)] = 3.67973$$
$$\overline{\log X = 4.13600}$$
$$X = 13677 \text{ ft.} = 4559 \text{ yds.}$$
$$\log [T(v) - T(V)] = 0.65610$$
$$\log \sec \varphi = 0.00665$$
$$\overline{\log T = 1.11902}$$
$$T = 13''.153$$
$$\log \left\{ I(v) - \frac{A(v) - A(V)}{S(v) - S(V)} \right\} = 9.20704$$
$$\log \sin 2\omega = 9.66331$$
$$2\omega = 27° 25' 30''$$
$$\omega = 13° 42' 45''$$

The value of ω computed by the more exact formula
is
$$\tan \omega = \frac{C}{2 \cos^2 \varphi} \left\{ I(v) - \frac{A(v) - A(V)}{S(v) - S(V)} \right\}$$
$$\omega = 13° \, 21' \, 30''$$
differing by $21'$ from the less approximate value.

We have found above
$$v = 873.8 \text{ f. s.}$$
but this is only an approximation. To determine its true value, that is, *its true value so far as the formulæ are concerned*, we should have
$$v = 873.8 \frac{\cos 10°}{\cos 13° \, 21' \, 30''} = 884.45 \text{ f. s.}$$

differing from the approximate value by about 10 feet.

Example 2.—" A 6-inch projectile leaves the gun at an angle of departure of $4°$, with an initial velocity of 2100 f. s.; $w = 64$ lbs., $d = 6$ inches. Find the range in horizontal plane through the muzzle of the gun, and time of flight." (" Exterior Ballistics," by Lieutenants Meigs and Ingersoll, U.S.N.)

We have (Table I.)
$$S(V) = 2024.8; \; A(V) = 20.57; \; I(V) = 0.02246; \; T(V) = 0.838$$
Taking $c = 1$, we have
$$C = \frac{64}{36}$$
Next we have
$$\frac{A(v) - 20.57}{S(v) - 2024.8} = \frac{36}{64} \sin 8° + I(V) = 0.10074$$
from which equation we readily find
$$v = 993.77 \text{ f. s.}$$
$$\therefore S(v) = 7801.8, \text{ and } T(v) = 5.051$$
$$X = C[7801.8 - 2024.8] = 10270 \text{ ft.}$$
$$T = \frac{C}{\cos \varphi} \left\{ 5.051 - 0.838 \right\} = 7''.51$$

PROBLEM 2.—*Given the angle of fall and terminal velocity, to determine the initial velocity, angle of projection, range, and time of flight.*

We have [equation (87)]

$$\frac{A(v) - A(V)}{S(v) - S(V)} = I(v) - \frac{\sin 2\omega}{C}$$

from which to calculate V by double position.

We may also determine V by the equation (see Problem 1)

$$T(V) = 2\,T(v_0) - T(v)$$

v_0 being found by the equation

$$I(v_0) = I(v) - \frac{\sin 2\omega}{C}$$

derived from (82).

Having found V by either method, φ, X, and T are computed by the equations

$$\sin 2\varphi = C\left\{\frac{A(v) - A(V)}{S(v) - S(V)} - I(V)\right\}$$

$$X = C\,[S(v) - S(V)]$$

$$T = \frac{C}{\cos\varphi}\left\{T(v) - T(V)\right\}$$

Example 1.—Given $d = 4.5$ inches; $w = 35$ lbs.; $\omega = 15°$, and $v = 772.74$ f. s.; to determine φ, X, and T.

It will be found that we have the following equation from which to find V:

$$\frac{2058.17 - A(V)}{11633.6 - S(V)} = 0.26807$$

For the first trial assume $V = 1500$, and, substituting in the first member of the above equation, it reduces it to 0.26691, which is too small by $0.00116 = e_1$. Next make $V = 1480$, and we shall find that the first member now becomes too great by $0.00140 = e_2$; then

$$256 : 20 :: 116 : 9.1$$

The correct value of V is therefore $1500 - 9.1 = 1490.9$ f. s., from which are easily found

$$\varphi = 9° 51'; \quad X = 12440 \text{ ft.}; \quad T = 12''.72.$$

Example 2.—" In attacking a place with curved fire it was required to drop shell into the place with an angle of descent of 12°, and terminal velocity of 600 f. s., using the 8-inch howitzer and a projectile of 180 lbs.; find the requisite position of the battery, and the requisite elevation and charge of powder." *

Here $d = 8$ inches; $w = 180$ lbs.; $v = 600$ f. s., and $\omega = 12°$; to find X, V, and φ. We have

$$\log \sin 2\omega = 9.60931$$
$$\log C = 0.44909$$
$$\log 0.14462 = 9.16022$$
$$I(v) = 1.15929$$
$$I(v_0) = 1.01467 \qquad v_0 = 630.85 \text{ f. s.}$$

whence we find

$$T(V) = 2 \times 14.396 - 15.779 = 13.012$$
$$V = 665.1 \text{ f. s.}$$
$$S(v) = 15926.6$$
$$S(V) = 14178.9$$
$$\log 1747.7 = 3.24247$$
$$\log X = 3.69156$$
$$X = 4915 \text{ ft.} = 1638 \text{ yds.}$$
$$I(v_0) = 1.01467$$
$$I(V) = 0.87708$$
$$\log 0.13759 = 9.13859$$
$$\log \sin 2\varphi = 9.58768$$
$$2\varphi = 22° 46' \qquad \varphi = 11° 23'$$

PROBLEM 3.—*Given the range and initial velocity, to determine the other elements of the trajectory.*

This is by far the most important of the ballistic problems, and it happens, fortunately, to be one of those most easily solved by Siacci's formulæ.

For the terminal velocity we have

$$S(v) = S(V) + \frac{X}{C}$$

---

* Prof. A. G. Greenhill in "Proceedings Royal Artillery Institution," No. 2, vol. xiii. page 79.

and then, with V and v known, all the other elements can be computed by formulæ already considered.

Example 1.—Find the elevation required for a range of 2000 yards with the 16-pdr. M. L. R. gun, the muzzle velocity being 1355 f. s.; find also the time of flight and angle of descent.

Here $d = 3.6$; $w = 16$; $\log C = 0.09152$; $V = 1355$, and $X = 6000$.

Answer :
$$\varphi = 4° 41'$$
$$T = 5''.91$$
$$\omega = 6° 13'$$

Example 2.—Compute a range table for the 8-inch rifle (converted), up to 15000 ft.

We have for chilled shot, $w = 183$ lbs.; $d = 8$ in. (whence $\log C = 0.45627$), and $V = 1404$ f. s. First take from Table I. the following numbers, which are to be used in all the calculations :

$$S(V) = 4858.5, \; A(V) = 162.23, \; I(V) = 0.08595, \; T(V) = 2.500$$

The remainder of the work may be concisely tabulated as follows :

X ft.	$\dfrac{X}{C}$	$S(v)$	v	$A(v)$	$I(v)$	$T(v)$
1500	524.59	5383.1	1303.0	212.04	0.10442	2.884
3000	1049.2	5907.7	1212.8	272.28	.12579	3.305
4500	1573.8	6432.3	1134.3	344.60	.15038	3.753
6000	2098.4	6956.9	1 69.2	430.79	.17826	4.230
7500	2622.9	7481.4	1019.2	532.14	.20929	4.732
9000	3147.5	8006.0	978.8	650.68	.24314	5.257
10500	3672.1	8530.6	942.5	787.72	.27973	5.804
12000	4196.7	9055.2	908.8	944.68	.31914	6.371
13500	4721.3	9579.8	877.4	1123.07	.36148	6.959
15000	5245.9	10104.4	848.1	1324.47	.40684	7.567

The numbers in the first column are the ranges for which the elements of the trajectory are to be computed. The numbers in the second column are simple multiples of the first number in the column. Adding $S(V)$ to the numbers in the second column gives those in the third column, and with these we take from Table I. the values of v, and at the same time those of $A(v)$, $I(v)$, and $T(v)$.

The time of flight, angle of departure, and angle of fall are then computed by the following formulæ:

$$\cdot T = \frac{C}{\cos \varphi} \left\{ T(v) - T(V) \right\}$$

and

$$\sin 2\varphi = C \left\{ \frac{A(v) - A(V)}{S(v) - S(V)} - I(V) \right\}$$

$$\tan \omega = \frac{C}{2 \cos^2 \varphi} \left\{ I(v) - \frac{A(v) - A(V)}{S(v) - S(V)} \right\}$$

Lastly, the values of v, tabulated above, are to be multiplied by $\cos \varphi \sec \omega$ to obtain the correct striking velocities.

In our example the results are as follows:

X yds	ϕ	ω	v f. s.	T
500	$0° \ 44'$	$0° \ 47'$	1303	$1''.10$
1000	$1° \ 33'$	$1° \ 43'$	1213	$2''.30$
1500	$2° \ 27'$	$2° \ 50'$	1135	$3''.59$
2000	$3° \ 27'$	$4° \ 08'$	1070	$4''.96$
2500	$4° \ 32'$	$5° \ 38'$	1021	$6''.40$
3000	$5° \ 43'$	$7° \ 14'$	982	$7''.92$
3500	$6° \ 59'$	$9° \ 01'$	947	$9''.52$
4000	$8° \ 21'$	$10° \ 58'$	916	$11''.19$
4500	$9° \ 49'$	$13° \ 06'$	888	$12''.94$
5000	$11° \ 24'$	$15° \ 25'$	862	$14''.78$

By interpolation, using first and second differences, the interval between successive values of the argument (X) may be reduced from 500 yards to 100 yards.

Example 3.—Given $d = 20.93$ cm.; $w = 140$ kg.; $V = 521$ m. s.; $\delta_{\prime} = 1.206$; $\delta = 1.233$; $X = 4097$ m.; angle of jump $= 8'$; required the angle of elevation $= \varphi - 8'$, the angle of fall, the striking velocity, and the time of flight.*

Making the ballistic coefficient $(c) = 0.907$, we have for

---

* " Ballistische Formeln von Mayevski nach Siacci. Für Elevationen unter 15 Grad," Essen, Fried. Krupp'sche Buchdruckerei, 1883, page 22. Also quoted by Siacci in " Rivista di Artiglieria e Genio," vol. ii. page 414, who solves the example, using Mayevski's table.

computing C in English units, when d is expressed in centimetres and w in kilogrammes, the following expression:

$$C = [1.1953743]\frac{\delta_{\prime}}{\delta}\frac{w}{d^2}$$

The following are the results obtained by experiment, by Mayevski's calculations, by Siacci's calculations, and by Table I. of this work:

	T	Angle of Elevation.	Angle of Fall.	Striking Velocity. f. s.
By experiment	9″.7	5° 30′		
Mayevski...	9″.6	5° 32′	7° 16′	1176
Siacci......	9″.675	5° 31′		
Table I.....	9″.66	5° 29′ 30″	7° 17′	1169

Example 4.—Given $d = 24$ cm.; $w = 215$ kg.; $V = 529$ m. s. $= 1735.6$ f. s.; required the angle of departure for each of the horizontal ranges contained in the first column of the following table:

Horizontal Range. m	$\frac{\delta_{\prime}}{\delta}$	ϕ Computed by Table I.	Observed value of ϕ	Values of ϕ computed by	
				Mayevski's Table.	Hojel's Table.
2026	0.9569	2° 17′	2° 19′	2° 18′	2° 14′
3000	0.9407	3° 36′	3° 41′	3° 37′	3° 35′
4000	0.9756	5° 5′	5° 10′	5° 6′	5° 5′
5964	0.9560	8° 41′	8° 35′	8° 44′	8° 44′
7600	0.9461	12° 31′	12° 5′	12° 31′	12° 32′

The data in the first, second, and fourth columns are taken from Krupp's Bulletin, No. 56 (February, 1885), page 4. The values of φ in the third column were computed by Siacci's method, using Table I. of this work. In the last two columns are given the values of φ computed by Siacci's method with Mayevski's and Hojel's tables respectively.

PROBLEM 4.—*With a given initial velocity, required the angle*

16

of projection necessary to cause a projectile to pass through a given point.

Let x and y be the co-ordinates of the given point. Then from (83) and (84) we have

$$S(v) = \frac{X}{C} + S(V)$$

and

$$\tan \varphi = \frac{y}{x} + \frac{C}{2 \cos^2 \varphi} \left\{ \frac{A(v) - A(V)}{S(v) - S(V)} - I(V) \right\}$$

Example.—An 8-inch service projectile is fired with an initial velocity of 1404 f. s. from a point 33 feet above the water; find the necessary angle of projection to attain a range on the water of 3000 yards.

Here $d = 8$, $w = 180$, $V = 1404$, $x = 9000$ ft., and $y = -33$ ft.

We have

$$S(v) = \frac{64}{180} \times 9000 + 4858.5 = 8058.5$$

$$\therefore v = 975.07$$

In calculating $\tan \varphi$ we will, at first, omit the factor $\cos^2 \varphi$ in the second member.

$$\therefore \tan \varphi = -\frac{33}{9000} + \frac{180}{128} \left\{ \frac{663.56 - 162.23}{8058.5 - 4858.5} - 0.08595 \right\}$$

$$= -0.00367 + 0.09945 = 0.09578$$

Therefore the approximate value of φ is $5° 28'$. Completing the calculation by introducing $\cos^2 \varphi$ we have

$$\varphi = 5° 31'$$

which needs no further correction.

PROBLEM 5.—*Given the initial and terminal velocities, to calculate the trajectory.*

For the solution of this problem we have the following equations:

$$\sin 2\varphi = C \left\{ \frac{A(v) - A(V)}{S(v) - S(V)} - I(V) \right\}$$

$$\sin 2\omega = C \left\{ I(v) - \frac{A(v) - A(V)}{S(v) - S(V)} \right\}$$

$$X = C[S(v) - S(V)]$$

$$T = \frac{C}{\cos \varphi} \left\{ T(v) - T(V) \right\}$$

Example.—In experimenting with the 15-inch S. B. gun, it is desired to place a target at such a distance from the gun that the projectile (solid shot weighing 450 lbs.) shall have a velocity of 1000 f. s. when it reaches the target, and this without diminishing the muzzle velocity, which is 1534 f. s. What is the required distance and the angle of projection?

We readily find, using Table II.,

and
$$\varphi = 2° \ 33'$$
$$X = 4678 \ \text{ft.}$$

CORRECTION FOR VARIATION IN THE DENSITY OF THE AIR.

The ballistic coefficient (C) is determined by the equation

$$C = \frac{w}{c \, d^2} \frac{\delta_{,}}{\delta}$$

in which $\delta_{,}$ is the adopted standard density of the air, and δ the density at the time of firing.

In computing Tables I. and II. the value of $\delta_{,}$ was taken as the weight, in grains, of a cubic foot of air at a temperature of 62° F. and a pressure of 30 inches of mercury. According to Bashforth we have

$$\delta_{,} = 534.22 \ \text{grs.}$$

For any other temperature (t), and barometric pressure (b), we may determine the value of δ near enough for most practical purposes by the following simple equation:

$$\delta = \frac{20.212 \ b}{1 + .002178 \ t}$$

Correction for Altitude.—When a projectile is fired at such an angle of projection as to reach a great altitude in its flight, the value of δ, determined as above, will be too great. We may calculate δ approximately, in this case, as follows:

If δ' is the density of the air at the height y above the surface of the earth, we shall have

$$\delta' = \delta \, c^{-\frac{y}{\lambda}}$$

where λ is the height of a homogeneous atmosphere of the density δ, which would exert a pressure equal to that of the actual atmosphere.*

The factor $\dfrac{\delta_{\prime}}{\delta}$ becomes, therefore, $\dfrac{\delta_{\prime}}{\delta} e^{\frac{y}{\lambda}}$; and C must be multiplied by this if we wish to take into account the diminution of density due to the height of the projectile, taking for y a mean value for the arc of the trajectory which we are computing.

The following table gives the values of $e^{\frac{y}{\lambda}}$ for every 100 feet from $y=0$ to $y=10{,}000$ feet. In the computation λ was assumed to be 27800 feet, which is its approximate value for a temperature of 15° C. and barometer at $0^m.75$. The table is substantially the same as that given by Bashforth ("Motion of Projectiles," page 103), but in a more convenient form.

y	0	100	200	300	400	500	600	700	800	900
0	1.0000	0036	0072	0108	0145	0181	0218	0255	0292	0329
1000	1.0366	0403	0441	0479	0516	0554	0592	0631	0669	0707
2000	1.0746	0785	0824	0863	0902	0941	0981	1020	1060	1100
3000	1.1140	1180	1220	1260	1301	1341	1382	1423	1464	1506
4000	1.1547	1589	1630	1672	1714	1756	1799	1841	1884	1927
5000	1.1970	2013	2057	2100	2144	2187	2231	2276	2320	2364
6000	1.2409	2454	2499	2544	2589	2634	2679	2725	2771	2817
7000	1.2863	2909	2956	3003	3049	3096	3144	3191	3239	3286
8000	1.3334	3382	3431	3479	3528	3576	3625	3675	3724	3773
9000	1.3823	3873	3923	3973	4023	4074	4125	4176	4227	4278

* Chauvenet's "Practical Astronomy," vol. i. page 138.

BALLISTIC TABLES.

THE term "Ballistic Table" was applied by Siacci to the tabulated values of the functions $S(v)$, $A(v)$, $I(v)$, and $T(v)$. Table I. gives the values of these functions for oblong projectiles having ogival heads struck with radii of $1\frac{1}{2}$ calibers. It is based upon the experiments of Bashforth, and was calculated by the formulæ developed in the preceding pages.

The table extends from $v = 2800$ to $v = 400$, which limits are extensive enough for the solution of nearly all practical problems of exterior ballistics. It may occasionally happen in mortar practice that the horizontal velocity ($v \cos \varphi$) may be less than 400 (as in problem 4, Chapter V.) In such cases we may employ the formulæ by which this part of the table was computed, viz.:

$$S(v) = 124466.4 - [4.5918330] \log v$$

$$A(v) = [9.6655206]\frac{1}{v^2} + [4.1438598] \log v - 45916.40$$

$$I(v) = [5.7369333]\frac{1}{v^2} - 0.356474$$

$$T(v) = [4.2296173]\frac{1}{v} - 12.4999$$

Example 1.—Let $d = 8$ in., $w = 180$ lbs., $V = 700$ f. s., and $\varphi = 60°$. Find v when $\vartheta = -60°$.

We have from (33)

$$I(u) = \frac{4\,(60°)}{C} + I(U)$$

and $U = 700 \cos 60° = 350$, which is below the limit of

the table. The operation may be concisely arranged as follows:

$$
\begin{aligned}
\text{const. } \log &= 5.7369333 \\
2 \log U &= 5.0881360 \\
\hline
0.6487973 &= \log 4.45448 \\
(60) &= 2.39053
\end{aligned}
$$

$$
\begin{aligned}
\log 4 \, (60^\circ) &= 0.9805542 \\
\log C &= 0.4490925 \\
\hline
0.5314617 &= \log 3.39987 \\
0.8951103 &= \log 7.85435 \\
\hline
2)4.8418230 \\
\hline
2\ 4209115 &= \log 263.6
\end{aligned}
$$

$$\therefore v = 263.6 \times 2 = 527.2 \text{ f. s.}$$

Example 2.—Given $S(v) = 25496.8$, to find v. We proceed as follows:

$$
\begin{aligned}
124466.4 \\
25496.8 \\
\hline
\log 98969.6 &= 4.9954886 \\
\text{const. } \log &= 4.5918330 \\
\hline
\log (\log v) &= 0.4036556 \\
\therefore \log v &= 2.53312 \\
\therefore v &= 341.3
\end{aligned}
$$

Table II. is the ballistic table for spherical projectiles, and extends from $v = 2000$ to $v = 450$. It is based upon the Russian experiments discussed in Chapter II., and is believed to be the only ballistic table for spherical projectiles yet published.

Table III. is abridged from Didion's " Traité de Balistique."

Formulæ for Interpolation.—To find the value of $f(v)$ when v lies between v_1 and v_2, two consecutive values of v, in Tables I. and II. Let $v_1 - v_2 = h$. Then, if d_1 and d_2

are the first and second differences of the function, we shall have, since $f(v)$ increases while v decreases,

$$f(v) = f(v_1) + \frac{v_1 - v}{h} d_1 - \frac{v_1 - v}{h}\left(1 - \frac{v_1 - v}{h}\right)\frac{d_2}{2}$$

by means of which $f(v)$ can be computed. Conversely, if $f(v)$ is given, and our object is to find v, we have

$$\frac{v_1 - v}{h} d_1 = f(v) - f(v_1) + \frac{v_1 - v}{h}\left(1 - \frac{v_1 - v}{h}\right)\frac{d_2}{2}$$

In using this last formula, first compute $\dfrac{v_1 - v}{h}$ by omitting the second term of the second member (which is usually very small), and then supply this term, using the approximate value of $\dfrac{v_1 - v}{h}$ already found.

If the second differences are too small to be taken into account, the above formulæ become

$$f(v) = f(v_1) + \frac{v_1 - v}{h} d_1$$

and

$$v = v_1 - \frac{h}{d_1}\left(f(v) - f(v_1)\right)$$

which expresses the ordinary rules of proportional parts.

Example 1.—Find from Table I. $S(v)$ when $v = 1432.6$. We have $v_1 = 1435$, $f(v_1) = 4704.8$, $h = 5$, and $d_1 = 24.6$.

$$\therefore S(v) = 4704.8 + \frac{1435 - 1432.6}{5} \times 24.6 = 4716.6$$

Example 2.—Given $A(v) = 229.89$, to find v. Here $v_1 = 1274$, $f(v_1) = 229.29$, $d_1 = 1.25$, and $h = 2$.

$$\therefore v = 1274 - \frac{2}{1.25}(229.89 - 229.29) = 1273.04$$

Example 3.—Find from Table II. $A(v)$ when $v = 517.8$.

We have $v_1 = 520$, $A(v_1) = 3755.9$, $h = 5$, $d_1 = 158.2$, and $d_2 = 7.8$.

$$\therefore A(v) = 3755.9 + \frac{2.2}{5} \times 158.2 - \frac{2.2}{5}\left(1 - \frac{2.2}{5}\right)\frac{7.8}{2}$$
$$= 3755.9 + 69.60 - 0.96 = 3824.5$$

Example 4.—Find from Table III. the value of (ϑ) when $\vartheta = 54° 32'$. Here $\vartheta_1 = 54° 20'$, $(\vartheta_1) = 1.76191$, $h = 20'$, $d_1 = .02971$, $d_2 = .00074$.

$$\therefore (\vartheta) = 1.76191 + 0.6 \times 0.02971 - 0.6 \times 0.4 \times 0.00037$$
$$= 1.76191 + 0.01783 - 0.00009 = 1.77965$$

TABLE I.

Ballistic Table for Ogival-Headed Projectiles.

v	$S(v)$	Diff.	$A(v)$	Diff.	$I(v)$	Diff.	$T'(v)$	Diff.
2800	000.0	1268	0.00	7	0.00000	106	0.000	46
2750	126.8	1292	0.07	21	0.00106	112	0.046	47
2700	256.0	1315	0.28	36	0.00218	118	0.093	49
2650	387.5	1341	0.64	54	0.00336	125	0.142	51
2600	521.6	1367	1.18	71	0.00461	133	0.193	53
2550	658.3	1393	1.89	93	0.00594	140	0.246	56
2500	797.6	1422	2.82	115	0.00734	149	0.302	57
2450	939.8	1452	3.97	140	0.00883	160	0.359	60
2400	1085.0	1481	5.37	166	0.01043	169	0.419	62
2350	1233.1	1514	7.03	197	0.01212	180	0.481	65
2300	1384.5	1547	9.00	231	0.01392	192	0.546	68
2250	1539.2	1582	11.31	266	0.01584	205	0.614	72
2200	1697.4	321	13.97	58	0.01789	43	0.686	14
2190	1729.5	322	14.55	60	0.01832	44	0.700	15
2180	1761.7	323	15.15	62	0.01876	44	0.715	15
2170	1794.0	325	15.77	63	0.01920	44	0.730	15
2160	1826.5	327	16.40	65	0.01964	46	0.745	15
2150	1859.2	328	17.05	67	0.02010	46	0.760	15
2140	1892.0	329	17.72	68	0.02056	46	0.775	16
2130	1924.9	331	18.40	70	0.02102	47	0.791	15
2120	1958.0	333	19.10	73	0.02149	48	0.806	16
2110	1991.3	335	19.83	74	0.02197	49	0.822	16
2100	2024.8	336	20.57	76	0.02246	49	0.838	16
2090	2058.4	337	21.33	79	0.02295	50	0.854	16
2080	2092.1	339	22.12	80	0.02345	51	0.870	16
2070	2126.0	341	22.92	82	0.02396	51	0.886	17
2060	2160.1	343	23.74	85	0.02447	52	0.903	17
2050	2194.4	344	24.59	87	0.02499	53	0.920	17
2040	2228.8	346	25.46	89	0.02552	54	0.937	17
2030	2263.4	348	26.35	91	0.02606	54	0.954	17
2020	2298.2	349	27.26	94	0.02660	55	0.971	17
2010	2333.1	351	28.20	96	0.02715	57	0.988	17
2000	2368.2	353	29.16	98	0.02772	57	1.005	18

5

TABLE I.—CONTINUED.

v	$S(v)$	Diff.	$A(v)$	Diff.	$I(v)$	Diff.	$T(v)$	Diff.
1990	2403.5	355	30.14	101	0.02829	57	1.023	18
1980	2439.0	356	31.15	104	0.02886	59	1.041	18
1970	2474.6	358	32.19	107	0.02945	60	1.059	18
1960	2510.4	360	33.26	109	0.03005	61	1.077	19
1950	2546.4	362	34.35	113	0.03066	61	1.096	18
1940	2582.6	363	35.48	115	0.03127	62	1.114	19
1930	2618.9	366	36.63	118	0.03189	64	1.133	19
1920	2655.5	367	37.81	121	0.03253	65	1.152	19
1910	2692.2	370	39.02	124	0.03318	65	1.171	20
1900	2729.2	371	40.26	127	0.03383	67	1.191	19
1890	2766.3	374	41.53	130	0.03450	67	1.210	20
1880	2803.7	375	42.83	133	0.03517	69	1.230	20
1870	2841.2	377	44.16	137	0.03586	70	1.250	20
1860	2878.9	380	45.53	140	0.03656	71	1.270	21
1850	2916.9	382	46.93	143	0.03727	72	1.291	20
1840	2955.1	383	48.36	147	0.03799	73	1.311	21
1830	2993.4	386	49.83	151	0.03872	74	1.332	21
1820	3032.0	388	51.34	155	0.03946	76	1.353	22
1810	3070.8	390	52.89	158	0.04022	77	1.375	21
1800	3109.8	392	54.47	162	0.04099	78	1.396	22
1790	3149.0	394	56.09	167	0.04177	80	1.418	22
1780	3188.4	396	57.76	171	0.04257	81	1.440	23
1770	3228.0	399	59.47	174	0.04338	82	1.463	22
1760	3267.9	401	61.21	179	0.04420	84	1.485	23
1750	3308.0	403	63.00	183	0.04504	85	1.508	23
1740	3348.3	406	64.83	188	0.04589	87	1.531	24
1730	3388.9	409	66.71	193	0.04676	88	1.555	23
1720	3429.8	410	68.64	197	0.04764	90	1.578	24
1710	3470.8	413	70.61	202	0.04854	91	1.602	24
1700	3512.1	415	72.63	207	0.04945	93	1.626	25
1690	3553.6	418	74.70	213	0.05038	95	1.651	25
1680	3595.4	420	76.83	218	0.05133	96	1.676	25
1670	3637.4	423	79.01	223	0.05229	98	1.701	25
1660	3679.7	425	81.24	228	0.05327	100	1.726	26
1650	3722.2	428	83.52	234	0.05427	102	1.752	26
1640	3765.0	430	85.86	241	0.05529	103	1.778	26

TABLE I.—CONTINUED.

v	$S(v)$	Diff.	$A(v)$	Diff.	$I(v)$	Diff.	$T(v)$	Diff.
1630	3808.0	433	88.27	246	0.05632	106	1.804	27
1620	3851.3	436	90.73	252	0.05738	107	1.831	27
1610	3894.9	438	93.25	259	0.05845	110	1.858	27
1600	3938.7	220	95.84	132	0.05955	55	1.885	14
1595	3960.7	221	97.16	133	0.06010	56	1.899	14
1590	3982.8	222	98.49	135	0.06066	57	1.913	14
1585	4005.0	223	99.84	137	0.06123	57	1.927	14
1580	4027.3	223	101.21	139	0.06180	58	1.941	14
1575	4049.6	224	102.60	140	0.06238	58	1.955	14
1570	4072.0	224	104.00	142	0.06296	59	1.969	14
1565	4094.4	225	105.42	144	0.06355	59	1.983	15
1560	4116.9	226	106.86	146	0.06414	60	1.998	14
1555	4139.5	227	108.32	147	0.06474	60	2.012	15
1550	4162.2	228	109.79	150	0.06534	61	2.027	15
1545	4185.0	228	111.29	151	0.06595	62	2.042	15
1540	4207.8	229	112.80	153	0.06657	− 62	2.057	15
1535	4230.7	229	114.33	155	0.06719	63	2.072	14
1530	4253.6	231	115.88	157	0.06782	64	2.086	15
1525	4276.7	231	117.45	159	0.06846	64	2.101	16
1520	4299.8	232	119.04	161	0.06910	65	2.117	15
1515	4323.0	232	120.65	163	0.06975	65	2.132	15
1510	4346.2	234	122.28	165	0.07040	66	2.147	15
1505	4369.6	234	123.93	167	0.07106	67	2.162	16
1500	4393.0	235	125.60	169	0.07173	68	2.178	16
1495	4416.5	236	127.29	172	0.07241	68	2.194	16
1490	4440.1	237	129.01	174	0.07309	69	2.210	16
1485	4463.8	237	130.75	175	0.07378	69	2.226	16
1480	4487.5	238	132.50	178	0.07447	70	2.242	16
1475	4511.3	239	134.28	181	0.07517	71	2.258	16
1470	4535.2	240	136.09	183	0.07588	72	2.274	16
1465	4559.2	240	137.92	185	0.07660	72	2.290	17
1460	4583.2	242	139.77	188	0.07732	73	2.307	16
1455	4607.4	242	141.65	189	0.07805	74	2.323	17
1450	4631.6	243	143.54	193	0.07879	75	2.340	17
1445	4655.9	244	145.47	195	0.07954	75	2.357	17
1440	4680.3	245	147.42	197	0.08029	76	2.374	17

TABLE I.—CONTINUED.

v	$S(v)$	Diff.	$A(v)$	Diff.	$I(v)$	Diff.	$T(v)$	Diff.
1435	4704.8	246	149.39	200	0.08105	77	2.391	17
1430	4729.4	247	151.39	203	0.08182	78	2.408	17
1425	4754.1	247	153.42	205	0.08260	78	2.425	18
1420	4778.8	248	155.47	208	0.08338	80	2.443	17
1415	4803.6	249	157.55	211	0.08418	81	2.460	18
1410	4828.5	250	159.66	214	0.08498	81	2.478	18
1405	4853.5	251	161.80	216	0.08579	82	2.496	18
1400	4878.6	252	163.96	219	0.08661	83	2.514	18
1395	4903.8	253	166.15	222	0.08744	84	2.532	18
1390	4929.1	254	168.37	225	0.08828	85	2.550	18
1385	4954.5	254	170.62	228	0.08913	86	2.568	19
1380	4979.9	256	172.90	231	0.08999	87	2.587	18
1375	5005.5	256	175.21	234	0.09086	87	2.605	19
1370	5031.1	257	177.55	237	0.09173	89	2.624	19
1365	5056.8	258	179.92	241	0.09262	89	2.643	19
1360	5082.6	260	182.33	243	0.09351	91	2.662	19
1355	5108.6	260	184.76	247	0.09442	91	2.681	19
1350	5134.6	261	187.23	250	0.09533	93	2.700	19
1345	5160.7	262	189.73	254	0.09626	94	2.719	20
1340	5186.9	263	192.27	257	0.09719	94	2.739	19
1335	5213.2	263	194.84	260	0.09813	95	2.758	20
1330	5239.5	263	197.44	262	0.09908	96	2.778	20
1325	5265.8	262	200.06	263	0.10004	97	2.798	20
1320	5292.0	106	202.69	107	0.10101	39	2.818	8
1318	5302.6	106	203.76	108	0.10140	39	2.826	8
1316	5313.2	106	204.84	108	0.10179	40	2.834	8
1314	5323.8	107	205.92	109	0.10219	40	2.842	8
1312	5334.5	107	207.01	110	0.10259	40	2.850	8
1310	5345.2	107	208.11	111	0.10299	40	2.858	8
1308	5355.9	108	209.22	111	0.10339	41	2.866	9
1306	5366.7	108	210.33	112	0.10380	41	2.875	8
1304	5377.5	108	211.45	113	0.10421	41	2.883	9
1302	5388.3	109	212.58	114	0.10462	41	2.892	8
1300	5399.2	109	213.72	115	0.10503	41	2.900	8
1298	5410.1	109	214.87	115	0.10544	42	2.908	9
1296	5421.0	110	216.02	117	0.10586	42	2.917	8

TABLE I.—Continued.

v	$S(v)$	Diff.	$A(v)$	Diff.	$I(v)$	Diff.	$T(v)$	Diff.
1294	5432.0	110	217.19	117	0.10628	42	2.925	9
1292	5443.0	110	218.36	118	0.10670	43	2.934	8
1290	5454.0	111	219.54	119	0.10713	43	2.942	8
1288	5465.1	111	220.73	120	0.10756	43	2.950	9
1286	5476.2	111	221.93	120	0.10799	43	2.959	9
1284	5487.3	112	223.13	122	0.10842	44	2.968	9
1282	5498.5	112	224.35	122	0.10886	44	2.977	8
1280	5509.7	113	225.57	123	0.10930	44	2.985	9
1278	5521.0	113	226.80	124	0.10974	45	2.994	9
1276	5532.3	113	228.04	125	0.11019	45	3.003	9
1274	5543.6	113	229.29	125	0.11064	45	3.012	9
1272	5554.9	114	230.54	127	0.11109	45	3.021	9
1270	5566.3	114	231.81	127	0.11154	46	3.030	9
1268	5577.7	114	233.08	129	0.11200	46	3.039	9
1266	5589.1	115	234.37	129	0.11246	46	3.048	9
1264	5600.6	115	235.66	131	0.11292	46	3.057	9
1262	5612.1	116	236.97	131	0.11338	47	3.066	9
1260	5623.7	116	238.28	132	0.11385	47	3.075	9
1258	5635.3	117	239.60	134	0.11432	47	3.084	10
1256	5647.0	116	240.94	134	0.11479	48	3.094	9
1254	5658.6	117	242.28	136	0.11527	48	3.103	10
1252	5670.3	118	243.64	136	0.11575	48	3.113	9
1250	5682.1	118	245.00	137	0.11623	48	3.122	9
1248	5693.9	118	246.37	139	0.11671	49	3.131	10
1246	5705.7	119	247.76	139	0.11720	49	3.141	9
1244	5717.6	119	249.15	140	0.11769	50	3.150	10
1242	5729.5	119	250.55	142	0.11819	50	3.160	9
1240	5741.4	120	251.97	142	0.11869	50	3.169	10
1238	5753.4	120	253.39	144	0.11919	50	3.179	10
1236	5765.4	121	254.83	144	0.11969	51	3.189	9
1234	5777.5	121	256.27	146	0.12020	51	3.198	10
1232	5789.6	121	257.73	147	0.12071	52	3.208	10
1230	5801.7	122	259.20	148	0.12123	52	3.218	10

TABLE I.—CONTINUED.

v	$S(v)$	Diff.	$A(v)$	Diff.	$I(v)$	Diff.	$T(v)$	Diff.
1228	5813.9	122	260.68	149	0.12175	52	3.228	10
1226	5826.1	123	262.17	150	0.12227	53	3.238	10
1224	5838.4	123	263.67	151	0.12280	53	3.248	10
1222	5850.7	123	265.18	153	0.12333	53	3.258	10
1220	5863.0	124	266.71	153	0.12386	53	3.268	10
1218	5875.4	124	268.24	155	0.12439	54	3.278	10
1216	5887.8	125	269.79	156	0.12493	54	3.288	11
1214	5900.3	125	271.35	157	0.12547	55	3.299	10
1212	5912.8	125	272.92	159	0.12602	55	3.309	10
1210	5925.3	126	274.51	160	0.12657	55	3.319	10
1208	5937.9	126	276.11	161	0.12712	56	3.329	11
1206	5950.5	127	277.72	162	0.12768	56	3.340	10
1204	5963.2	127	279.34	163	0.12824	57	3.350	11
1202	5975.9	127	280.97	165	0.12881	57	3.361	10
1200	5988.6	128	282.62	166	0.12938	57	3.371	11
1198	6001.4	128	284.28	167	0.12995	58	3.382	11
1196	6014.2	129	285.95	168	0.13053	58	3.393	11
1194	6027.1	129	287.63	170	0.13111	58	3.404	11
1192	6040.0	130	289.33	171	0.13169	59	3.415	11
1190	6053.0	130	291.04	172	0.13228	59	3.426	11
1188	6066.0	131	292.76	174	0.13287	60	3.437	11
1186	6079.1	131	294.50	175	0.13347	60	3.448	11
1184	6092.2	131	296.25	177	0.13407	60	3.459	11
1182	6105.3	132	298.02	178	0.13467	61	3.470	11
1180	6118.5	132	299.80	179	0.13528	61	3.481	11
1178	6131.7	133	301.59	181	0.13589	62	3.492	12
1176	6145.0	133	303.40	182	0.13651	62	3.504	11
1174	6158.3	134	305.22	184	0.13713	63	3.515	12
1172	6171.7	134	307.06	185	0.13776	63	3.527	11
1170	6185.1	135	308.91	186	0.13839	63	3.538	12
1168	6198.6	135	310.77	188	0.13902	64	3.550	11
1166	6212.1	135	312.65	190	0.13966	64	3.561	12
1164	6225.6	136	314.55	191	0.14030	65	3.573	11

TABLE I.—CONTINUED.

v	$S(v)$	Diff.	$A(v)$	Diff.	$I(v)$	Diff.	$T(v)$	Diff.
1162	6239.2	136	316.46	193	0.14095	65	3.584	12
1160	6252.8	69	318.39	97	0.14160	32	3.596	6
1159	6259.7	69	319.36	98	0.14192	33	3.602	6
1158	6266.6	68	320.34	98	0.14225	33	3.608	6
1157	6273.4	69	321.32	98	0.14258	33	3.614	6
1156	6280.3	69	322.30	98	0.14291	33	3.620	6
1155	6287.2	69	323.28	99	0.14324	34	3.626	6
1154	6294.1	69	324.27	99	0.14358	33	3.632	6
1153	6301.0	69	325.26	100	0.14391	34	3.638	6
1152	6307.9	69	326.26	100	0.14425	33	3.644	6
1151	6314.8	70	327.26	101	0.14458	34	3.650	6
1150	6321.8	70	328.27	101	0.14492	34	3.656	6
1149	6328.8	69	329.28	101	0.14526	34	3.662	6
1148	6335.7	70	330.29	102	0.14560	34	3.668	6
1147	6342.7	70	331.31	102	0.14594	34	3.674	6
1146	6349.7	70	332.33	103	0.14628	34	3.680	6
1145	6356.7	70	333.36	103	0.14662	35	3.686	7
1144	6363.7	70	334.39	104	0.14697	34	3.693	6
1143	6370.7	71	335.43	104	0.14731	35	3.699	6
1142	6377.8	70	336.47	104	0.14766	35	3.705	6
1141	6384.8	71	337.51	105	0.14801	35	3.711	6
1140	6391.9	71	338.56	105	0.14836	35	3.717	6
1139	6399.0	71	339.61	106	0.14871	35	3.723	7
1138	6406.1	71	340.67	106	0.14906	36	3.730	6
1137	6413.2	71	341.73	106	0.14942	35	3.736	6
1136	6420.3	71	342.79	107	0.14977	36	3.742	6
1135	6427.4	72	343.86	108	0.15013	36	3.748	7
1134	6434.6	71	344.94	108	0.15049	36	3.755	6
1133	6441.7	72	346.02	108	0.15085	36	3.761	6
1132	6448.9	72	347.10	109	0.15121	36	3.767	7
1131	6456.1	72	348.19	109	0.15157	36	3.774	6
1130	6463.3	71	349.28	110	0.15193	36	3.780	6
1129	6470.4	72	350.38	109	0.15229	36	3.786	7

TABLE I.—CONTINUED.

v	$S(v)$	Diff.	$A(v)$	Diff.	$I(v)$	Diff.	$T(v)$	Diff.
1128	6477.6	72	351.47	110	0.15265	37	3.793	6
1127	6484.8	73	352.57	111	0.15302	36	3.799	7
1126	6492.1	72	353.68	111	0.15338	37	3.806	6
1125	6499.3	73	354.79	111	0.15375	37	3.812	6
1124	6506.6	73	355.90	113	0.15412	37	3.818	7
1123	6513.9	73	357.03	113	0.15449	38	3.825	6
1122	6521.2	74	358.16	114	0.15487	37	3.831	7
1121	6528.6	74	359.30	115	0.15524	38	3.838	6
1120	6536.0	74	360.45	115	0.15562	38	3.844	7
1119	6543.4	74	361.60	116	0.15600	38	3.851	7
1118	6550.8	75	362.76	116	0.15638	38	3.858	6
1117	6558.3	75	363.92	117	0.15676	39	3.864	7
1116	6565.8	75	365.09	119	0.15715	39	3.871	7
1115	6573.3	75	366.28	119	0.15754	39	3.878	7
1114	6580.8	76	367.47	120	0.15793	39	3.885	7
1113	6588.4	76	368.67	121	0.15832	40	3.892	6
1112	6596.0	77	369.88	121	0.15872	40	3.898	7
1111	6603.7	77	371.09	123	0.15912	40	3.905	7
1110	6611.4	77	372.32	123	0.15952	41	3.912	7
1109	6619.1	78	373.55	124	0.15993	40	3.919	7
1108	6626.9	78	374.79	125	0.16033	41	3.926	7
1107	6634.7	78	376.04	126	0.16074	41	3.933	7
1106	6642.5	78	377.30	127	0.16115	42	3.940	7
1105	6650.3	79	378.57	128	0.16157	41	3.947	8
1104	6658.2	80	379.85	129	0.16198	42	3.955	7
1103	6666.2	79	381.14	130	0.16240	42	3.962	7
1102	6674.1	80	382.44	131	0.16282	43	3.969	7
1101	6682.1	81	383.75	131	0.16325	42	3.976	7
1100	6690.2	81	385.06	132	0.16367	43	3.983	8
1099	6698.3	81	386.38	133	0.16410	43	3.991	7
1098	6706.4	81	387.71	135	0.16453	44	3.998	8
1097	6714.5	82	389.06	135	0.16497	44	4.006	7
1096	6722.7	83	390.41	137	0.16541	44	4.013	8

TABLE I.—CONTINUED.

v	$S(v)$	Diff.	$A(v)$	Diff.	$I(v)$	Diff.	$T(v)$	Diff.
1095	6731.0	82	391.78	137	0.16585	44	4.021	8
1094	6739.2	83	393.15	138	0.16629	45	4.029	7
1093	6747.5	84	394.53	140	0.16674	45	4.036	8
1092	6755.9	84	395.93	141	0.16719	45	4.044	7
1091	6764.3	84	397.34	141	0.16764	46	4.051	8
1090	6772.7	85	398.75	142	0.16810	46	4.059	8
1089	6781.2	85	400.17	143	0.16856	46	4.067	8
1088	6789.7	85	401.60	145	0.16902	46	4.075	8
1087	6798.2	86	403.05	145	0.16948	47	4.083	8
1086	6806.8	86	404.50	147	0.16995	47	4.091	7
1085	6815.4	87	405.97	148	0.17042	47	4.098	8
1084	6824.1	87	407.45	149	0.17089	48	4.106	8
1083	6832.8	87	408.94	150	0.17137	48	4.114	8
1082	6841.5	88	410.44	151	0.17185	48	4.122	8
1081	6850.3	88	411.95	152	0.17233	49	4.130	8
1080	6859.1	88	413.47	153	0.17282	49	4.138	8
1079	6867.9	89	415.00	154	0.17331	49	4.146	9
1078	6876.8	90	416.54	156	0.17380	49	4.155	8
1077	6885.8	89	418.10	156	0.17429	50	4.163	9
1076	6894.7	90	419.66	158	0.17479	50	4.172	8
1075	6903.7	91	421.24	159	0.17529	51	4.180	9
1074	6912.8	91	422.83	161	0.17580	51	4.189	8
1073	6921.9	92	424.44	162	0.17631	51	4.197	9
1072	6931.1	92	426.06	163	0.17682	51	4.206	8
1071	6940.3	92	427.69	164	0.17733	52	4.214	9
1070	6949.5	93	429.33	165	0.17785	52	4.223	9
1069	6958.8	93	430.98	166	0.17837	53	4.232	9
1068	6968.1	94	432.64	168	0.17890	53	4.241	9
1067	6977.5	94	434.32	169	0.17943	53	4.250	9
1066	6986.9	94	436.01	171	0.17996	53	4.259	9
1065	6996.3	95	437.72	172	0.18049	54	4.268	9
1064	7005.8	96	439.44	173	0.18103	55	4.277	9
1063	7015.4	96	441.17	175	0.18158	55	4.286	9

TABLE I.—Continued.

v	$S(v)$	Diff.	$A(v)$	Diff.	$I(v)$	Diff.	$T(v)$	Diff.
1062	7025.0	96	442.92	176	0.18213	55	4.295	9
1061	7034.6	97	444.68	177	0.18268	55	4.304	9
1060	7044.3	97	446.45	178	0.18323	56	4.313	9
1059	7054.0	98	448.23	180	0.18379	56	4.322	10
1058	7063.8	98	450.03	181	0.18435	56	4.332	9
1057	7073.6	99	451.84	182	0.18491	57	4.341	9
1056	7083.5	99	453.66	184	0.18548	57	4.350	10
1055	7093.4	100	455.50	186	0.18605	58	4.360	9
1054	7103.4	100	457.36	187	0.18663	58	4.369	9
1053	7113.4	100	459.23	189	0.18721	58	4.378	9
1052	7123.4	101	461.12	190	0.18779	59	4.387	10
1051	7133.5	102	463.02	192	0.18838	59	4.397	9
1050	7143.7	102	464.94	193	0.18897	59	4.406	10
1049	7153.9	102	466.87	194	0.18956	60	4.416	10
1048	7164.1	103	468.81	196	0.19016	61	4.426	10
1047	7174.4	103	470.77	197	0.19077	61	4.436	10
1046	7184.7	104	472.74	199	0.19138	61	4.446	9
1045	7195.1	105	474.73	201	0.19199	61	4.455	10
1044	7205.6	105	476.74	203	0.19260	62	4.465	10
1043	7216.1	105	478.77	204	0.19322	63	4.475	10
1042	7226.6	106	480.81	206	0.19385	63	4.485	10
1041	7237.2	107	482.87	208	0.19448	63	4.495	10
1040	7247.9	107	484.95	209	0.19511	64	4.505	11
1039	7258.6	107	487.04	211	0.19575	64	4.516	10
1038	7269.3	108	489.15	213	0.19639	64	4.526	11
1037	7280.1	109	491.28	214	0.19703	65	4.537	10
1036	7291.0	109	493.42	216	0.19768	66	4.547	11
1035	7301.9	110	495.58	218	0.19834	66	4.558	11
1034	7312.9	110	497.76	219	0.19900	66	4.569	10
1033	7323.9	111	499.95	222	0.19966	67	4.579	11
1032	7335.0	111	502.17	223	0.20033	67	4.590	10
1031	7346.1	112	504.40	225	0.20100	68	4.600	11
1030	7357.3	112	506.65	226	0.20168	68	4.611	11

TABLE I.—CONTINUED.

v	$S(v)$	Diff.	$A(v)$	Diff.	$I(v)$	Diff.	$T(v)$	Diff.
1029	7368.5	113	508.91	229	0.20236	69	4.622	11
1028	7379.8	113	511.20	230	0.20305	69	4.633	12
1027	7391.1	114	513.50	232	0.20374	69	4.645	11
1026	7402.5	115	515.82	235	0.20443	70	4.656	11
1025	7414.0	115	518.17	237	0.20513	71	4.667	11
1024	7425.5	116	520.54	238	0.20584	71	4.678	11
1023	7437.1	116	522.92	240	0.20655	71	4.689	12
1022	7448.7	117	525.32	243	0.20726	72	4.701	11
1021	7460.4	117	527.75	245	0.20798	73	4.712	11
1020	7472.1	118	530.20	246	0.20871	73	4.723	12
1019	7483.9	118	532.66	248	0.20944	73	4.735	12
1018	7495.7	119	535.14	251	0.21017	74	4.747	12
1017	7507.6	120	537.65	252	0.21091	74	4.759	12
1016	7519.6	120	540.17	255	0.21165	75	4.771	11
1015	7531.6	121	542.72	258	0.21240	76	4.782	12
1014	7543.7	121	545.30	259	0.21316	76	4.794	12
1013	7555.8	122	547.89	262	0.21392	76	4.806	12
1012	7568.0	123	550.51	265	0.21468	77	4.818	12
1011	7580.3	123	553.16	266	0.21545	78	4.830	12
1010	7592.6	124	555.82	269	0.21623	78	4.842	13
1009	7605.0	124	558.51	272	0.21701	79	4.855	12
1008	7617.4	125	561.23	273	0.21780	79	4.867	13
1007	7629.9	126	563.96	275	0.21859	80	4.880	12
1006	7642.5	126	566.71	278	0.21939	80	4.892	13
1005	7655.1	127	569.49	280	0.22019	81	4.905	13
1004	7667.8	128	572.29	282	0.22100	82	4.918	12
1003	7680.6	128	575.11	285	0.22182	82	4.930	13
1002	7693.4	129	577.96	287	0.22264	83	4.943	12
1001	7706.3	130	580.83	289	0.22347	83	4.955	13
1000	7719.3	131	583.72	292	0.22430	84	4.968	13
999	7732.4	132	586.64	295	0.22514	85	4.981	14
998	7745.6	132	589.59	297	0.22599	85	4.995	13
997	7758.8	133	592.56	300	0.22684	86	5.008	14

TABLE I.—CONTINUED.

v	$S(v)$	Diff.	$A(v)$	Diff.	$I(v)$	Diff.	$T(v)$	Diff.
996	7772.1	133	595.56	303	0.22770	87	5.022	13
995	7785.4	133	598.59	306	0.22857	87	5.035	13
994	7798.7	134	601.65	309	0.22944	87	5.048	14
993	7812.1	134	604.74	311	0.23031	87	5.062	13
992	7825.5	135	607.85	314	0.23118	88	5.075	14
991	7839.0	135	610.99	317	0.23206	89	5.089	13
990	7852.5	136	614.16	317	0.23295	89	5.102	14
989	7866.1	136	617.33	319	0.23384	90	5.116	14
988	7879.7	137	620.52	321	0.23474	90	5.130	14
987	7893.4	137	623.73	323	0.23564	91	5.144	14
986	7907.1	137	626.96	325	0.23655	91	5.158	13
985	7920.8	137	630.21	327	0.23746	91	5.171	14
984	7934.5	138	633.48	329	0.23837	92	5.185	14
983	7948.3	138	636.77	331	0.23929	92	5.199	14
982	7962.1	138	640.08	333	0.24021	92	5.213	14
981	7975.9	139	643.41	335	0.24113	93	5.227	14
980	7989.8	139	646.76	336	0.24206	93	5.241	14
979	8003.7	139	650.12	339	0.24299	93	5.255	15
978	8017.6	139	653.51	341	0.24392	94	5.270	14
977	8031.5	140	656.92	343	0.24486	94	5.284	15
976	8045.5	140	660.35	345	0.24580	95	5.299	14
975	8059.5	140	663.80	346	0.24675	95	5.313	14
974	8073.5	141	667.26	349	0.24770	95	5.327	15
973	8087.6	141	670.75	351	0.24865	96	5.342	14
972	8101.7	141	674.26	354	0.24961	96	5.356	15
971	8115.8	141	677.80	355	0.25057	97	5.371	14
970	8129.9	142	681.35	357	0.25154	97	5.385	15
969	8144.1	142	684.92	359	0.25251	97	5.400	15
968	8158.3	142	688.51	361	0.25348	98	5.415	14
967	8172.5	143	692.12	363	0.25446	98	5.429	15
966	8186.8	143	695.75	366	0.25544	99	5.444	15
965	8201.1	143	699.41	368	0.25643	99	5.459	15
964	8215.4	144	703.09	370	0.25742	99	5.474	15

TABLE I.—CONTINUED.

v	$S(v)$	Diff.	$A(v)$	Diff.	$I(v)$	Diff.	$T(v)$	Diff.
963	8229.8	144	706.79	372	0.25841	100	5.489	14
962	8244.2	144	710.51	375	0.25941	100	5.503	15
961	8258.6	144	714.26	377	0.26041	101	.5.518	15
960	8273.0	144	718.03	378	0.26142	101	5.533	15
959	8287.4	145	721.81	381	0.26243	101	5.548	16
958	8301.9	145	.725.62	384	0.26344	102	5.564	15
957	8316.4	146	729.46	386	0.26446	103	5.579	15
956	8331.0	146	733.32	388	0.26549	103	5.594	15
955	8345.6	146	737.20	390	0.26652	103	5.609	16
954	8360.2	146	741.10	393	0.26755	103	5.625	15
953	8374.8	147	745.03	395	0.26858	104	5.640	15
952	8389.5	147	748.98	398	0.26962	105	5.655	16
951	8404.2	148	752.96	400	0.27067	105	5.671	15
950	8419.0	148	756.96	402	0.27172	105	5.686	16
949	8433.8	148	760.98	404	0.27277	106	5.702	16
948	8448.6	148	765.02	407	0.27383	106	5.718	15
947	8463.4	149	769.09	409	0.27489	107	.5.733	16
946	8478.3	149	773.18	412	0.27596	107	5.749	16
945	8493.2	149	777.30	415	0.27703	108	5.765	16
944	8508.1	150	781.45	417	0.27811	108	5.781	16
943	8523.1	150	785.62	420	0.27919	108	5.797	15
942	8538.1	150	789.82	422	0.28027	109	5.812	16
941	8553.1	151	794.04	425	0.28136	110	5.828	16
940	8568.2	151	798.29	427	0.28246	110	5.844	16
939	8583.3	151	802.56	429	0.28356	111	5.860	17
938	8598.4	152	806.85	432	0.28467	111	5.877	16
937	8613.6	152	811.17	435	0.28578	111	5.893	16
936	8628.8	152	815.52	437	0.28689	112	5.909	17
935	8644.0	152	819.89	441	0.28801	112	5.926	16
934	8659.2	153	824.30	443	0.28913	113	5.942	16
933	8674.5	153	828.73	445	0.29026	114	5.958	16
932	8689.8	154	833.18	449	0.29140	114	5.974	17
931	8705.2	154	837.67	451	0.29254	114	.5.991	16

TABLE I.—CONTINUED.

v	$S(v)$	Diff.	$A(v)$	Diff.	$I(v)$	Diff.	$T(v)$	Diff.
930	8720.6	154	842.18	453	0.29368	115	6.007	17
929	8736.0	155	846.71	456	0.29483	115	6.024	17
928	8751.5	155	851.27	459	0.29598	116	6.041	16
927	8767.0	155	855.86	462	0.29714	116	6.057	17
926	8782.5	155	860.48	465	0.29830	117	6.074	17
925	8798.0	156	865.13	468	0.29947	117	6.091	17
924	8813.6	156	869.81	470	0.30064	118	6.108	17
923	8829.2	157	874.51	474	0.30182	118	6.125	16
922	8844.9	157	879.25	477	0 30300	119	6.141	17
921	8860.6	157	884.02	479	0.30419	119	6.158	17
920	8876.3	157	888.81	482	0.30538	120	6.175	17
919	8892.0	158	893.63	485	0.30658	120	6.192	18
918	8907.8	159	898.48	488	0.30778	121	6.210	17
917	8923.7	158	903.36	491	0.30899	121	6.227	18
916	8939.5	159	908.27	494	0.31020	122	6.245	17
915	8955.4	159	913.21	497	0.31142	122	6.262	17
914	8971.3	160	918.18	501	0.31264	123	6.279	18
913	8987.3	160	923.19	503	0.31387	124	6.297	17
912	9003.3	160	928.22	506	0.31511	124	6.314	18
911	9019.3	161	933.28	509	0.31635	125	6.332	17
910	9035.4	161	938.37	513	0.31760	125	6.349	18
909	9051.5	161	943.50	515	0.31885	126	6.367	18
908	9067.6	162	948.65	519	0.32011	126	6.385	18
907	9083.8	162	953.84	522	0.32137	127	6.403	18
906	9100.0	162	959.06	525	0.32264	128	6.421	18
905	9116.2	163	964.31	529	0.32392	128	6.439	18
904	9132.5	163	969.60	532	0.32520	129	6.457	18
903	9148.8	164	974.92	535	0.32649	129	6.475	18
902	9165.2	164	980.27	538	0.32778	130	6.493	18
901	9181.6	164	985.65	541	0.32908	130	6.511	18
900	9198.0	165	991.06	545	0.33038	131	6.529	19
899	9214.5	165	996.51	548	0.33169	131	6.548	18
898	9231.0	165	1001.99	552	0.33300	132	6.566	19

TABLE I.—Continued.

v	$S(v)$	Diff.	$A(v)$	Diff.	$I(v)$	Diff.	$T(v)$	Diff.
897	9247.5	166	1007.51	555	0.33432	133	6.585	18
896	9264.1	166	1013.06	559	0.33565	133	6.603	19
895	9280.7	166	1018.65	562	0.33698	134	6.622	18
894	9297.3	167	1024.27	565	0.33832	134	6.640	19
893	9314.0	167	1029.92	569	0.33966	135	6.659	18
892	9330.7	168	1035.61	573	0.34101	136	6.677	19
891	9347.5	168	1041.34	576	0.34237	136	6.696	18
890	9364.3	168	1047.10	580	0.34373	137	6.714	19
889	9381.1	169	1052.90	583	0.34510	137	6.733	20
888	9398.0	169	1058.73	587	0.34647	138	6.753	19
887	9414.9	170	1064.60	592	0.34785	139	6.772	19
886	9431.9	170	1070.52	595	0.34924	139	6.791	20
885	9448.9	170	1076.47	598	0.35063	140	6.811	19
884	9465.9	171	1082.45	602	0.35203	141	6.830	19
883	9483.0	171	1088.47	606	0.35344	141	6.849	19
882	9500.1	171	1094.53	609	0.35485	142	6.868	20
881	9517.2	172	1100.62	613	0.35627	143	6.888	19
880	9534.4	172	1106.75	617	0.35770	143	6.907	20
879	9551.6	173	1112.92	621	0.35913	144	6.927	20
878	9568.9	173	1119.13	625	0.36057	145	6.947	19
877	9586.2	173	1125.38	629	0.36202	145	6.966	20
876	9603.5	174	1131.67	633	0.36347	146	6.986	20
875	9620.9	174	1138.00	637	0.36493	146	7.006	20
874	9638.3	175	1144.37	641	0.36639	147	7.026	20
873	9655.8	175	1150.78	645	0.36786	148	7.046	19
872	9673.3	175	1157.23	649	0.36934	149	7.065	20
871	9690.8	176	1163.72	653	0.37083	149	7.085	20
870	9708.4	176	1170.25	657	0.37232	150	7.105	21
869	9726.0	177	1176.82	662	0.37382	150	7.126	20
868	9743.7	177	1183.44	665	0.37532	151	7.146	21
867	9761.4	177	1190.09	670	0.37683	152	7.167	20
866	9779.1	178	1196.79	675	0.37835	153	7.187	21
865	9796.9	178	1203.54	678	0.37988	153	7.208	21

TABLE I.—Continued.

v	$S(v)$	Diff.	$A(v)$	Diff.	$I(v)$	Diff.	$T(v)$	Diff.
864	9814.7	179	1210.32	683	0.38141	154	7.229	20
863	9832.6	179	1217.15	687	0.38295	155	7.249	21
862	9850.5	179	1224.02	691	0.38450	156	7.270	20
861	9868.4	180	1230.93	696	0.38606	156	7.290	21
860	9886.4	180	1237.89	700	0.38762	157	7.311	21
859	9904.4	181	1244.89	705	0.38919	158	7.332	22
858	9922.5	181	1251.94	710	0.39077	158	7.354	21
857	9940.6	181	1259.04	714	0.39235	159	7.375	21
856	9958.7	182	1266.18	718	0.39394	160	7.396	22
855	9976.9	183	1273.36	723	0.39554	161	7.418	21
854	9995.2	183	1280.59	728	0.39715	162	7.439	21
853	10013.5	183	1287.87	732	0.39877	162	7.460	21
852	10031.8	184	1295.19	737	0.40039	163	7.481	22
851	10050.2	184	1302.56	742	0.40202	164	7.503	21
850	10068.6	185	1309.98	746	0.40366	164	7.524	22
849	10087.1	185	1317.44	752	0.40530	165	7.546	22
848	10105.6	185	1324.96	756	0.40695	166	7.568	22
847	10124.1	186	1332.52	761	0.40861	167	7.590	22
846	10142.7	186	1340.13	766	0.41028	168	7.612	23
845	10161.3	187	1347.79	771	0.41196	168	7.635	22
844	10180.0	188	1355.50	776	0.41364	169	7.657	22
843	10198.8	187	1363.26	781	0.41533	170	7.679	22
842	10217.5	188	1371.07	786	0.41703	171	7.701	22
841	10236.3	189	1378.93	791	0.41874	172	7.723	22
840	10255.2	189	1386.84	796	0.42046	172	7.745	23
839	10274.1	189	1394.80	802	0.42218	174	7.768	22
838	10293.0	190	1402.82	807	0.42392	174	7.790	23
837	10312.0	190	1410.89	812	0.42566	175	7.813	23
836	10331.0	191	1419.01	817	0.42741	176	7.836	22
835	10350.1	191	1427.18	823	0.42917	176	7.858	23
834	10369.2	192	1435.41	828	0.43093	178	7.881	23
833	10388.4	192	1443.69	833	0.43271	178	7.904	24
832	10407.6	193	1452.02	839	0.43449	180	7.928	23

TABLE I.—Continued.

v	$S(v)$	Diff.	$A(v)$	Diff.	$I(v)$	Diff.	$T(v)$	Diff.
831	10426.9	193	1460.41	844	0.43629	180	7.951	23
830	10446.2	194	1468.85	850	0.43809	181	7.974	23
829	10465.6	194	1477.35	855	0.43990	182	7.997	24
828	10485.0	194	1485.90	861	0.44172	182	8.021	23
827	10504.4	195	1494.51	867	0.44354	184	8.044	24
826	10523.9	195	1503.18	872	0.44538	184	8.068	23
825	10543.4	196	1511.90	879	0.44722	186	8.091	24
824	10563.0	197	1520.69	883	0.44908	186	8.115	24
823	10582.7	197	1529.52	890	c.45094	188	8.139	24
822	10602.4	197	1538.42	896	0.45282	188	8.163	24
821	10622.1	198	1547.38	901	0.45470	189	8.187	24
820	10641.9	198	1556.39	908	0.45659	190	8.211	24
819	10661.7	199	1565.47	914	0.45849	191	8.235	24
818	10681.6	200	1574.61	919	0.46040	191	8.259	25
817	10701.6	200	1583.80	925	0.46231	193	8.284	24
816	10721.6	200	1593.05	932	0.46424	194	8.308	25
815	10741.6	201	1602.37	938	0.46618	194	8.333	24
814	10761.7	201	1611.75	945	0.46812	196	8.357	25
813	10781.8	202	1621.20	950	0.47008	197	8.382	25
812	10802.0	202	1630.70	957	0.47205	197	8.407	25
811	10822.2	203	1640.27	963	0.47402	199	8.432	25
810	10842.5	203	1649.90	970	0.47601	199	8.457	25
809	10862.8	204	1659.60	976	0.47800	201	8.482	25
808	10883.2	204	1669.36	983	0.48001	201	8.507	26
807	10903.6	205	1679.19	989	0.48202	202	8.533	25
806	10924.1	205	1689.08	996	0.48404	204	8.558	26
805	10944.6	206	1699.04	1003	0.48608	204	8.584	26
804	10965.2	206	1709.07	1009	0.48812	206	8.610	25
803	10985.8	207	1719.16	1016	0.49018	207	8.635	26
802	11006.5	207	1729.32	1023	0.49225	207	8.661	26
801	11027.2	208	1739.55	1029	0.49432	209	8.687	26
800	11048.0	208	1749.84	1037	0.49641	209	8.713	26
799	11068.8	209	1760.21	1043	0.49850	211	8.739	26

TABLE I.—CONTINUED.

v	$S(v)$	Diff.	$A(v)$	Diff.	$I(v)$	Diff.	$T(v)$	Diff.
798	11089.7	210	1770.64	1051	0.50061	212	8.765	26
797	11110.7	210	1781.15	1057	0.50273	213	8.791	27
796	11131.7	210	1791.72	1065	0.50486	214	8.818	26
795	11152.7	211	1802.37	1073	0.50700	215	8.844	27
794	11173.8	212	1813.10	1079	0.50915	216	8.871	26
793	11195.0	212	1823.89	1087	0.51131	217	8.897	27
792	11216.2	213	1834.76	1094	0.51348	218	8.924	27
791	11237.5	213	1845.70	1101	0.51566	220	8.951	27
790	11258.8	215	1856.71	1116	0.51786	222	8.978	27
789	11280.3	215	1867.87	1121	0.52008	223	9.005	27
788	11301.8	216	1879.08	1128	0.52231	223	9.032	28
787	11323.4	216	1890.36	1134	0.52454	224	9.060	27
786	11345.0	216	1901.70	1141	0.52678	226	9.087	27
785	11366.6	216	1913.11	1146	0.52904	226	9.114	28
784	11388.2	216	1924.57	1153	0.53130	227	9.142	28
783	11409.8	217	1936.10	1160	0.53357	228	9.170	27
782	11431.5	218	1947.70	1166	0.53585	228	9.197	28
781	11453.3	217	1959.36	1172	0.53813	230	9.225	28
780	11475.0	218	1971.08	1179	0.54043	230	9.253	28
779	11496.8	218	1982.87	1185	0.54273	231	9.281	28
778	11518.6	218	1994.72	1192	0.54504	232	9.309	28
777	11540.4	218	2006.64	1198	0.54736	233	9.337	28
776	11562.2	219	2018.62	1206	0.54969	234	9.365	29
775	11584.1	219	2030.68	1212	0.55203	235	9.394	28
774	11606.0	219	2042.80	1218	0.55438	236	9.422	28
773	11627.9	220	2054.98	1226	0.55674	237	9.450	29
772	11649.9	220	2067.24	1232	0.55911	237	9.479	28
771	11671.9	220	2079.56	1239	0.56148	239	9.507	29
770	11693.9	221	2091.95	1246	0.56387	239	9.536	29
769	11716.0	220	2104.41	1253	0.56626	241	9.565	28
768	11738.0	221	2116.94	1260	0.56867	241	9.593	29
767	11760.1	222	2129.54	1267	0.57108	242	9.622	29
766	11782.3	222	2142.21	1274	0.57350	244	9.651	29

TABLE I.—Continued.

v	$S(v)$	Diff.	$A(v)$	Diff.	$I(v)$	Diff.	$T(v)$	Diff.
765	11804.5	222	2154.95	1281	0.57594	244	9.680	29
764	11826.7	222	2167.76	1288	0.57838	245	9.709	29
763	11848.9	222	2180.64	1295	0.58083	247	9.738	29
762	11871.1	223	2193.59	1303	0.58330	247	9.767	30
761	11893.4	223	2206.62	1309	0.58577	248	9.797	29
760	11915.7	223	2219.71	1317	0.58825	249	9.826	29
759	11938.0	224	2232.88	1324	0.59074	250	9.855	30
758	11960.4	224	2246.12	1332	0.59324	251	9.885	29
757	11982.8	225	2259.44	1339	0.59575	252	9.914	30
756	12005.3	224	2272.83	1347	0.59827	253	9.944	29
755	12027.7	225	2286.30	1354	0.60080	254	9.973	30
754	12050.2	226	2299.84	1361	0.60334	255	10.003	30
753	12072.8	225	2313.45	1369	0.60589	256	10.033	30
752	12095.3	226	2327.14	1377	0.60845	258	10.063	30
751	12117.9	226	2340.91	1384	0.61103	258	10.093	30
750	12140.5	226	2354.75	1392	0.61361	259	10.123	30
749	12163.1	227	2368.67	1399	0.61620	260	10.153	31
748	12185.8	227	2382.66	1408	0.61880	262	10.184	30
747	12208.5	227	2396.74	1415	0.62142	262	10.214	30
746	12231.2	227	2410.89	1423	0.62404	263	10.244	31
745	12253.9	228	2425.12	1432	0.62667	265	10.275	31
744	12276.7	229	2439.44	1439	0.62932	266	10.306	30
743	12299.6	228	2453.83	1447	0.63198	266	10.336	31
742	12322.4	229	2468.30	1456	0.63464	268	10.367	31
741	12345.3	229	2482.86	1463	0.63732	269	10.398	31
740	12368.2	229	2497.49	1472	0.64001	270	10.429	31
739	12391.1	230	2512.21	1480	0.64271	271	10.460	31
738	12414.1	230	2527.01	1488	0.64542	272	10.491	31
737	12437.1	230	2541.89	1497	0.64814	273	10.522	32
736	12460.1	231	2556.86	1505	0.65087	274	10.554	31
735	12483.2	231	2571.91	1513	0.65361	276	10.585	31
734	12506.3	231	2587.04	1521	0.65637	276	10.616	32
733	12529.4	232	2602.25	1530	0.65913	278	10.648	31

TABLE I.—Continued.

v	$S(v)$	Diff.	$A(v)$	Diff.	$I(v)$	Diff.	$T(v)$	Diff.
732	12552.6	232	2617.55	1539	0.66191	279	10.679	32
731	12575.8	232	2632.94	1547	0.66470	280	10.711	32
730	12599.0	233	2648.41	1556	0.66750	281	10.743	32
729	12622.3	233	2663.97	1564	0.67031	282	10.775	32
728	12645.6	233	2679.61	1573	0.67313	283	10.807	32
727	12668.9	234	2695.34	1582	0.67596	285	10.839	32
726	12692.3	233	2711.16	1591	0.67881	286	10.871	32
725	12715.6	234	2727.07	1600	0.68167	287	10.903	33
724	12739.0	235	2743.07	1609	0.68454	288	10.936	32
723	12762.5	235	2759.16	1617	0.68742	289	10.968	33
722	12786.0	235	2775.33	1627	0.69031	291	11.001	32
721	12809.5	236	2791.60	1636	0.69322	292	11.033	33
720	12833.1	236	2807.96	1645	0.69614	293	11.066	33
719	12856.7	236	2824.41	1655	0.69907	294	11.099	33
718	12880.3	236	2840.96	1664	0.70201	295	11.132	33
717	12903.9	237	2857.60	1673	0.70496	297	11.165	33
716	12927.6	237	2874.33	1682	0.70793	298	11.198	33
715	12951.3	238	2891.15	1692	0.71091	299	11.231	33
714	12975.1	238	2908.07	1701	0.71390	301	11.264	33
713	12998.9	238	2925.08	1711	0.71691	302	11.297	33
712	13022.7	238	2942.19	1720	0.71993	303	11.330	34
711	13046.5	239	2959.39	1730	0.72296	304	11.364	34
710	13070.4	239	2976.69	1740	0.72600	305	11.398	34
709	13094.3	240	2994.09	1749	0.72905	307	11.432	33
708	13118.3	240	3011.58	1759	0.73212	308	11.465	34
707	13142.3	240	3029.17	1769	0.73520	310	11.499	34
706	13166.3	240	3046.86	1780	0.73830	311	11.533	34
705	13190.3	241	3064.66	1789	0.74141	312	11.567	34
704	13214.4	241	3082.55	1799	0.74453	313	11.601	35
703	13238.5	242	3100.54	1810	0.74766	315	11.636	34
702	13262.7	242	3118.64	1820	0.75081	316	11.670	34
701	13286.9	242	3136.84	1830	0.75397	318	11.704	35
700	13311.1	242	3155.14	1841	0.75715	319	11.739	35

TABLE I.—CONTINUED.

v	$S(v)$	Diff.	$A(v)$	Diff.	$I(v)$	Diff.	$T(v)$	Diff.
699	13335.3	243	3173.55	1851	0.76034	320	11.774	35
698	13359.6	243	3192.06	1861	0.76354	321	11.809	35
697	13383.9	244	3210.67	1872	0.76675	3 3	11.844	35
696	13408.3	244	3229.39	1883	0.76998	324	11.879	35
695	13432.7	244	3248.22	1893	0.77322	326	11.914	35
694	13457.1	245	3267.15	1904	0.77648	327	11.949	35
693	13481.6	245	3286.19	1914	0.77975	329	11.984	36
692	13506.1	245	3305.33	1925	0.78304	330	12.020	35
691	13530.6	246	3324.58	1937	0.78634	332	12.055	36
690	13555.2	246	3343.95	1947	0.78966	333	12.091	35
689	13579.8	246	3363.42	1958	0.79299	334	12.126	36
688	13604.4	247	3383.00	1970	0.79633	336	12.162	36
687	13629.1	247	3402.70	1980	0.79969	337	12.198	36
686	13653.8	248	3422.50	1992	0.80306	339	12.234	36
685	13678.6	248	3442.42	2003	0.80645	340	12.270	36
684	13703.4	248	3462.45	2015	0.80985	342	12.306	36
683	13728.2	249	3482.60	2026	0.81327	343	12.342	37
682	13753.1	249	3502.86	2038	0.81670	345	12.379	36
681	13778.0	249	3523.24	2049	0.82015	347	12.415	37
680	13802.9	250	3543.73	2061	0.82362	348	12.452	37
679	13827.9	250	3564.34	2073	0.82710	349	12.489	37
678	13852.9	250	3585.07	2084	0.83059	351	12.526	37
677	13877.9	251	3605.91	2097	0.83410	352	12.563	37
676	13903 0	251	3626.88	2108	0.83762	354	12.600	37
675	13928.1	252	3647.96	2121	0.84116	356	12.637	38
674	13953.3	252	3669.17	2133	0.84472	357	12.675	37
673	13978.5	252	3690.50	2144	0.84829	359	12.712	38
672	14003.7	253	3711 94	2157	0.85188	361	12.750	37
671	14029.0	253	3733.51	2170	0.85549	362	12.787	38
670	14054.3	253	3755.21	2182	0.85911	363	12.825	38
669	14079.6	254	3777.03	2195	0.86274	365	12.863	38
668	14105.0	254	3798.98	2207	0.86639	367	12.901	38
667	14130.4	255	3821.05	2219	0.87006	369	12.939	38

TABLE I.—Continued.

v	$S(v)$	Diff.	$A(v)$	Diff.	$I(v)$	Diff.	$T(v)$	Diff.
666	14155.9	255	3843.24	2233	0.87375	370	12.977	38
665	14181.4	255	3865.57	2245	0.87745	372	13.015	38
664	14206.9	256	3888.02	2258	0.88117	373	13.053	39
663	14232.5	256	3910.60	2271	0.88490	376	13.092	38
662	14258.1	256	3933.31	2285	0.88866	377	13.130	39
661	14283.7	257	3956.16	2297	0.89243	379	13.169	39
660	14309.4	257	3979.13	2311	0.89622	380	13.208	39
659	14335.1	258	4002.24	2324	0.90002	382	13.247	39
658	14360.9	258	4025.48	2338	0.90384	384	13.286	40
657	14386.7	259	4048.86	2351	0.90768	385	13.326	39
656	14412.6	259	4072.37	2364	0.91153	388	13.365	39
655	14438.5	259	4096.01	2378	0.91541	389	13.404	40
654	14464.4	260	4119.79	2392	0.91930	391	13.444	40
653	14490.4	260	4143.71	2406	0.92321	394	13.484	40
652	14516.4	260	4167.77	2419	0.92715	395	13.524	40
651	14542.4	261	4191.96	2434	0.93110	396	13.564	40
650	14568.5	261	4216.30	2448	0.93506	398	13.604	40
649	14594.6	262	4240.78	2462	0.93904	400	13.644	40
648	14620.8	262	4265.40	2476	0.94304	402	13.684	41
647	14647.0	262	4290.16	2491	0.94706	404	13.725	41
646	14673.2	263	4315.07	2505	0.95110	406	13.766	40
645	14699.5	264	4340.12	2520	0.95516	407	13.806	41
644	14725.9	264	4365.32	2535	0.95923	410	13.847	41
643	14752.3	264	4390.67	2549	0.96333	412	13.888	41
642	14778.7	264	4416.16	2565	0.96745	413	13.929	42
641	14805.1	265	4441.81	2579	0.97158	416	13.971	41
640	14831.6	265	4467.60	2595	0.97574	417	14.012	41
639	14858.1	266	4493.55	2609	0.97991	419	14.053	42
638	14884.7	266	4519.64	2625	0.98410	421	14.095	42
637	14911.3	267	4545.89	2641	0.98831	423	14.137	42
636	14938.0	267	4572.30	2656	0.99254	426	14.179	42
635	14964.7	267	4598.86	2671	0.99680	427	14.221	42
634	14991.4	268	4625.57	2687	1.00107	429	14.263	42

TABLE I.—CONTINUED.

v	$S(v)$	Diff.	$A(v)$	Diff.	$I(v)$	Diff.	$T(v)$	Diff.
633	15018.2	268	4652.44	2703	1.00536	431	14.305	43
632	15045.0	269	4679.47	2718	1.00967	434	14.348	42
631	15071.9	269	4706.65	2735	1.01401	436	14.390	43
630	15098.8	270	4734.00	2751	1.01837	437	14.433	43
629	15125.8	270	4761.51	2767	1.02274	439	14.476	43
628	15152.8	270	4789.18	2784	1.02713	442	14.519	43
627	15179.8	271	4817.02	2800	1.03155	443	14.562	43
626	15206.9	271	4845.02	2816	1.03598	446	14.605	43
625	15234.0	272	4873.18	2833	1.04044	448	14.648	44
624	15261.2	272	4901.51	2849	1.04492	451	14.692	43
623	15288.4	273	4930.00	2867	1.04943	452	14.735	44
622	15315.7	273	4958.67	2883	1.05395	455	14.779	44
621	15343.0	273	4987.50	2901	1.05850	457	14.823	44
620	15370.3	274	5016.51	2918	1.06307	459	14.867	44
619	15397.7	274	5045.69	2935	1.06766	461	14.911	45
618	15425.1	275	5075.04	2953	1.07227	463	14.956	44
617	15452.6	275	5104.57	2970	1.07690	466	15.000	45
616	15480.1	276	5134.27	2988	1.08156	468	15.045	45
615	15507.7	276	5164.15	3006	1.08624	471	15.090	45
614	15535.3	277	5194.21	3023	1.09095	473	15.135	45
613	15563.0	277	5224.44	3042	1.09568	475	15.180	45
612	15590.7	277	5254.86	3060	1.10043	477	15.225	45
611	15618.4	278	5285.46	3078	1.10520	480	15.270	46
610	15646.2	278	5316.24	3097	1.11000	482	15.316	45
609	15674.0	279	5347.21	3115	1.11482	484	15.361	46
608	15701.9	279	5378.36	3135	1.11966	486	15.407	46
607	15729.8	280	5409.71	3153	1.12452	489	15.453	46
606	15757.8	280	5441.24	3171	1.12941	492	15.499	47
605	15785.8	281	5472.95	3191	1.13433	494	15.546	46
604	15813.9	281	5504.86	3210	1.13927	497	15.592	46
603	15842.0	281	5536.96	3230	1.14424	499	15.638	47
602	15870.1	282	5569.26	3249	1.14923	502	15.685	47
601	15898.3	283	5601.75	3268	1.15425	504	15.732	47

TABLE I.—CONTINUED.

v	$S(v)$	Diff.	$A(v)$	Diff.	$I(v)$	Diff.	$T(v)$	Diff.
600	15926.6	283	5634.43	3288	1.15929	506	15.779	47
599	15954.9	283	5667.31	3309	1.16435	509	15.826	47
598	15983.2	284	5700.40	3329	1.16944	512	15.873	48
597	16011.6	285	5733.69	3349	1.17456	514	15.921	47
596	16040.1	285	5767.18	3369	1.17970	517	15.968	48
595	16068.6	285	5800.87	3389	1.18487	519	16.016	48
594	16097.1	286	5834.76	3409	1.19006	522	16.064	49
593	16125.7	287	5868.85	3431	1.19528	525	16.113	48
592	16154.4	287	5903.16	3451	1.20053	527	16.161	48
591	16183.1	287	5937.67	3472	1.20580	530	16.209	49
590	16211.8	288	5972.39	3493	1.21110	533	16.258	49
589	16240.6	288	6007.32	3515	1.21643	535	16.307	49
588	16269.4	289	6042.47	3536	1.22178	538	16.356	49
587	16298.3	289	6077.83	3558	1.22716	541	16.405	49
586	16327.2	290	6113.41	3579	1.23257	544	16.454	50
585	16356.2	290	6149.20	3602	1.23801	547	16.504	49
584	16385.2	291	6185.22	3624	1.24348	549	16.553	50
583	16414.3	291	6221.46	3646	1.24897	552	16.603	50
582	16443.4	292	6257.92	3669	1.25449	555	16.653	51
581	16472.6	292	6294.61	3691	1.26004	558	16.704	50
580	16501.8	293	6331.52	3714	1.26562	561	16.754	51
579	16531.1	293	6368.66	3735	1.27123	564	16.805	50
578	16560.4	294	6406.01	3762	1.27687	566	16.855	51
577	16589.8	294	6443.63	3783	1.28253	570	16.906	52
576	16619.2	295	6481.46	3806	1.28823	573	16.958	51
575	16648.7	295	6519.52	3830	1.29396	575	17.009	51
574	16678.2	296	6557.82	3854	1.29971	579	17.060	52
573	16707.8	296	6596.36	3878	1.30550	581	17.112	52
572	16737.4	297	6635.14	3902	1.31131	585	17.164	52
571	16767.1	298	6674.16	3926	1.31716	588	17.216	52
570	16796.9	298	6713.42	3951	1.32304	591	17.268	52
569	16826.7	299	6752.93	3975	1.32895	594	17.320	53
568	16856.6	299	6792.68	4000	1.33489	597	17.373	52

TABLE I.—CONTINUED.

v	$S(v)$	Diff.	$A(v)$	Diff.	$I(v)$	Diff.	$T(v)$	Diff.
567	16886.5	299	6832.68	4025	1.34086	600	17.425	53
566	16916.4	300	6872.93	4050	1.34686	604	17.478	53
565	16946.4	301	6913.43	4075	1.35290	607	17.531	53
564	16976.5	301	6954.18	4101	1.35897	610	17.584	54
563	17006.6	302	6995.19	4127	1.36507	613	17.638	53
562	17036.8	302	7036.46	4153	1.37120	616	17.691	54
561	17067.0	303	7077.99	4179	1.37736	620	17.745	54
560	17097.3	303	7119.78	4205	1.38356	623	17.799	54
559	17127.6	304	7161.83	4232	1.38979	627	17.853	55
558	17158.0	304	7204.15	4258	1.39606	630	17.908	54
557	17188.4	305	7246.73	4285	1.40236	633	17.962	55
556	17218.9	305	7289.58	4313	1.40869	637	18.017	55
555	17249.4	306	7332.71	4340	1.41506	640	18.072	55
554	17280.0	307	7376.11	4367	1.42146	643	18.127	56
553	17310.7	307	7419.78	4396	1.42789	647	18.183	55
552	17341.4	308	7463.74	4423	1.43436	651	18.238	56
551	17372.2	308	7507.97	4451	1.44087	654	18.294	56
550	17403.0	309	7552.48	4480	1.44741	658	18.350	56
549	17433.9	309	7597.28	4508	1.45399	661	18.406	56
548	17464.8	310	7642.36	4537	1.46060	665	18.462	57
547	17495.8	310	7687.73	4566	1.46725	669	18.519	57
546	17526.8	311	7733.39	4595	1.47394	672	18.576	57
545	17557.9	312	7779.34	4624	1.48066	676	18.633	57
544	17589.1	312	7825.58	4654	1.48742	680	18.690	57
543	17620.3	313	7872.12	4684	1.49422	684	18.747	58
542	17651.6	313	7918.96	4716	1.50106	687	18.805	58
541	17682.9	314	7966.12	4743	1.50793	691	18.863	58
540	17714.3	315	8013.55	4775	1.51484	695	18.921	58
539	17745.8	315	8061.30	4806	1.52179	699	18.979	59
538	17777.3	316	8109.36	4837	1.52878	703	19.038	58
537	17808.9	316	8157.73	4868	1.53581	706	19.096	59
536	17840.5	317	8206.41	4900	1.54287	711	19.155	60
535	17872.2	317	8255.41	4932	1.54998	715	19.215	59

TABLE I.—CONTINUED.

v	$S(v)$	Diff	$A'(v)$	Diff.	$I(v)$	Diff.	$T(v)$	Diff.
534	17903.9	318	8304.73	4963	1.55713	718	19.274	60
533	17935.7	319	8354.36	4996	1.56431	723	19.334	60
532	17967.6	319	8404.32	5029	1.57154	727	19.394	60
531	17999.5	320	8454.61	5061	1.57881	731	19.454	60
530	18031.5	320	8505.22	5094	1.58612	735	19.514	60
529	18063.5	321	8556.16	5128	1.59347	739	19.574	61
528	18095.6	322	8607.44	5162	1.60086	744	19.635	61
527	18127.8	322	8659.06	5195	1.60830	748	19.696	61
526	18160.0	323	8711.01	5229	1.61578	752	19.757	62
525	18192.3	324	8763.30	5264	1.62330	756	19.819	62
524	18224.7	324	8815.94	5298	1.63086	761	19.881	62
523	18257.1	325	8868.92	5333	1.63847	765	19.943	62
522	18289.6	325	8922.25	5368	1.64612	769	20.005	62
521	18322.1	326	8975.93	5404	1.65381	774	20.067	63
520	18354.7	327	9029.97	5439	1.66155	778	20.130	63
519	18387.4	327	9084.36	5475	1.66933	783	20.193	63
518	18420.1	328	9139.11	5512	1.67716	788	20.256	63
517	18452.9	328	9194.23	5548	1.68504	792	20.319	64
516	18485.7	329	9249.71	5585	1.69296	796	20.383	64
515	18518.6	330	9305.56	5623	1.70092	802	20.447	64
514	18551.6	331	9361.79	5660	1.70894	806	20.511	64
513	18584.7	331	9418.39	5699	1.71700	810	20.575	65
512	18617.8	332	9475.38	5736	1.72510	816	20.640	65
511	18651.0	332	9532.74	5775	1.73326	820	20.705	65
510	18684.2	333	9590.49	5813	1.74146	825	20.770	65
509	18717.5	334	9648.62	5853	1.74971	830	20.835	66
508	18750.9	334	9707.15	5891	1.75801	835	20.901	66
507	18784.3	335	9766.06	5932	1.76636	840	20.967	66
506	18817.8	336	9825.38	5971	1.77476	845	21.033	66
505	18851.4	336	9885.09	6012	1.78321	850	21.099	67
504	18885.0	337	9945.21	6053	1.79171	855	21.166	67
503	18918.7	338	10005.74	6093	1.80026	860	21.233	67
502	18952.5	338	10066.67	6134	1.80886	865	21.300	67

TABLE I.—CONTINUED.

v	$S(v)$	Diff.	$A(v)$	Diff.	$I(v)$	Diff.	$T(v)$	Diff.
501	18986.3	339	10128.01	6177	1.81751	871	21.367	68
500	19020.2	340	10189.78	6219	1.82622	876	21.435	68
499	19054.2	340	10251.9	626	1.83498	881	21.503	69
498	19088.2	341	10314.5	631	1.84379	886	21.572	69
497	19122.3	341	10377.6	634	1.85265	892	21.641	69
496	19156.4	342	10441.0	639	1.86157	897	21.710	69
495	19190.6	343	10504.9	644	1.87054	903	21.779	69
494	19224.9	344	10569.3	648	1.87957	908	21.848	70
493	19259.3	345	10634.1	652	1.88865	913	21.918	70
492	19293.8	345	10699.3	657	1.89778	919	21.988	70
491	19328.3	346	10765.0	661	1.90697	925	22.058	70
490	19362.9	347	10831.1	665	1.91622	930	22.128	71
489	19397.6	347	10897.6	671	1.92552	936	22.199	71
488	19432.3	348	10964.7	675	1.93488	942	22.270	71
487	19467.1	349	11032.2	679	1.94430	948	22.341	72
486	19502.0	349	11100.1	685	1.95378	954	22.413	72
485	19536.9	351	11168.6	689	1.96332	960	22.485	72
484	19572.0	351	11237.5	695	1.97292	966	22.557	73
483	19607.1	351	11307.0	699	1.98258	972	22.630	73
482	19642.2	353	11376.9	703	1.99230	977	22.703	73
481	19677.5	353	11447.2	709	2.00207	983	22.776	73
480	19712.8	354	11518.1	713	2.01190	990	22.849	74
479	19748.2	354	11589.4	719	2.02180	996	22.923	74
478	19783.6	355	11661.3	724	2.03176	1003	22.997	74
477	19819.1	356	11733.7	729	2.04179	1009	23.071	75
476	19854.7	357	11806.6	734	2.05188	1015	23.146	75
475	19890.4	358	11880.0	739	2.06203	1022	23.221	75
474	19926.2	358	11953.9	745	2.07225	1028	23.296	76
473	19962.0	359	12028.4	750	2.08253	1035	23.372	76
472	19997.9	360	12103.4	755	2.09288	1041	23.448	76
471	20033.9	361	12178.9	760	2.10329	1047	23.524	77
470	20070.0	362	12254.9	766	2.11376	1054	23.601	77
469	20106.2	362	12331.5	771	2.12430	1061	23.678	77

TABLE I.—CONTINUED.

v	$S(v)$	Diff.	$A(v)$	Diff.	$I(v)$	Diff.	$T(v)$	Diff.
468	20142.4	363	12408.6	777	2.13491	1068	23.755	78
467	20178.7	363	12486.3	783	2.14559	1076	23.833	78
466	20215.0	365	12564.6	788	2.15635	1082	23.911	78
465	20251.5	365	12643.4	794	2.16717	1089	23.989	79
464	20288.0	367	12722.8	799	2.17806	1096	24.068	79
463	20324.7	367	12802.7	805	2.18902	1104	24.147	79
462	20361.4	367	12883.2	811	2.20006	1110	24.226	80
461	20398.1	369	12964.3	816	2.21116	1117	24.306	80
460	20435.0	369	13045.9	822	2.22233	1124	24.386	80
459	20471.9	370	13128.1	829	2.23357	1132	24.466	81
458	20508.9	371	13211.0	834	2.24489	1140	24.547	81
457	20546.0	371	13294.4	841	2.25629	1147	24.628	82
456	20583.1	373	13378.5	848	2.26776	1155	24.710	82
455	20620.4	373	13463.3	853	2.27931	1163	24.792	82
454	20657.7	374	13548.6	859	2.29094	1171	24.874	82
453	20695.1	375	13634.5	866	2.30265	1178	24.956	83
452	20732.6	376	13721.1	872	2.31443	1185	25.039	83
451	20770.2	377	13808.3	878	2.32628	1193	25.122	84
450	20807.9	377	13896.1	885	2.33821	1201	25.206	84
449	20845.6	378	13984.6	891	2.35022	1210	25.290	84
448	20883.4	380	14073.7	898	2.36232	1218	25.374	85
447	20921.4	380	14163.5	905	2.37450	1226	25.459	85
446	20959.4	380	14254.0	911	2.38676	1235	25.544	85
445	20997.4	382	14345.1	919	2.39911	1243	25.629	86
444	21035.6	383	14437.0	925	2.41154	1251	25.715	86
443	21073.9	383	14529.5	932	2.42405	1260	25.801	87
442	21112.2	385	14622.7	939	2.43665	1268	25.888	87
441	21150.7	385	14716.6	946	2.44933	1276	25.975	87
440	21189.2	386	14811.2	953	2.46209	1285	26.062	88
439	21227.8	387	14906.5	960	2.47494	1294	26.150	88
438	21266.5	388	15002.5	968	2.48788	1303	26.238	89
437	21305.3	389	15099.3	975	2.50091	1313	26.327	89
436	21344.2	389	15196.8	982	2.51404	1322	26.416	89

TABLE I.—CONTINUED.

v	$S(v)$	Diff.	$A(v)$	Diff.	$I(v)$	Diff.	$T(v)$	Diff.
435	21383.1	391	15295.0	990	2.52726	1331	26.505	90
434	21422.2	392	15394.0	997	2.54057	1340	26.595	90
433	21461.4	392	15493.7	1005	2.55397	1349	26.685	91
432	21500.6	394	15594.2	1012	2.56746	1358	26.776	91
431	21540.0	394	15695.4	1019	2.58104	1367	26.867	92
430	21579.4	395	15797.3	1027	2.59471	1377	26.959	92
429	21618.9	396	15900.0	1035	2.60848	1387	27.051	92
428	21658.5	397	16003.5	1044	2.62235	1397	27.143	93
427	21698.2	398	16107.9	1052	2.63632	1407	27.236	93
426	21738.0	399	16213.1	1060	2.65039	1417	27.329	94
425	21777.9	399	16319.1	1068	2.66456	1427	27.423	94
424	21817.8	401	16425.9	1076	2.67883	1437	27.517	95
423	21857.9	402	16533.5	1084	2.69320	1447	27.612	95
422	21898.1	403	16641.9	1093	2.70767	1458	27.707	96
421	21938.4	403	16751.2	1101	2.72225	1467	27.803	96
420	21978.7	404	16861.3	1109	2.73692	1477	27.899	96
419	22019.1	405	16972.2	1119	2.75169	1489	27.995	97
418	22059.6	406	17084.1	1127	2.76658	1500	28.092	97
417	22100.2	407	17196.8	1137	2.78158	1510	28.189	98
416	22140.9	409	17310.5	1145	2.79668	1522	28.287	98
415	22181.8	409	17425.0	1155	2.81190	1533	28.385	99
414	22222.7	410	17540.5	1163	2.82723	1544	28.484	99
413	22263.7	411	17656.8	1173	2.84267	1555	28.583	100
412	22304.8	413	17774.1	1181	2.85822	1566	28.683	100
411	22346.1	413	17892.2	1191	2.87388	1577	28.783	101
410	22387.4	414	18011.3	1200	2.88965	1589	28.884	101
409	22428.8	416	18131.3	1211	2.90554	1601	28.985	102
408	22470.4	416	18252.4	1220	2.92155	1613	29.087	102
407	22512.0	417	18374.4	1230	2.93768	1625	29.189	103
406	22553.7	419	18497.4	1240	2.95393	1637	29.292	103
405	22595.6	419	18621.4	1250	2.97030	1649	29.395	104
404	22637.5	421	18746.4	1259	2.98679	1662	29.499	104
403	22679.6	422	18872.3	1270	3.00341	1674	29.603	105

TABLE I.—Continued.

v	$S(v)$	Diff.	$A(v)$	Diff.	$I(v)$	Diff.	$T(v)$	Diff.
402	22721.8	422	18999.3	1280	3.02015	1686	29.708	105
401	22764.0	424	19127.3	1289	3.03701	1698	29.813	106
400	22806.4	424	19256.2	1300	3.05399	1710	29.919	106

34

TABLE II.

For Spherical Projectiles.

v	$S(v)$	Diff.	$A(v)$	Diff.	$I(v)$	Diff.	$T(v)$	Diff.
2000	0	25	0.00	1	0.00000	40	0.000	12
1990	25	24	0.01	1	00040	40	0.012	13
1980	49	25	0.02	2	00080	41	0.025	12
1970	74	25	0.04	4	0.00121	42	0.037	13
1960	99	25	0.08	5	00163	42	0.050	13
1950	124	26	0.13	5	00205	43	0.063	13
1940	150	25	0.18	7	0.00248	44	0.076	13
1930	175	26	0.25	8	00292	44	0.089	13
1920	201	25	0.33	9	00336	45	0.102	14
1910	226	26	0.42	11	0.00381	46	0.116	13
1900	252	26	0.53	12	00427	46	0.129	14
1890	278	26	0.65	13	00473	47	0.143	14
1880	304	26	0.78	14	0.00520	48	0.157	14
1870	330	27	0.92	15	00568	49	0.171	14
1860	357	26	1.07	17	00617	49	0.185	14
1850	383	26	1.24	19	0.00666	50	0.199	15
1840	409	27	1.43	20	00716	51	0.214	14
1830	436	27	1.63	21	00767	52	0.228	15
1820	463	27	1.84	23	0.00819	53	0.243	15
1810	490	27	2.07	24	00872	54	0.258	15
1800	517	28	2.31	26	00926	55	0.273	15
1790	545	27	2.57	27	0.00981	55	0.288	16
1780	572	28	2.84	30	01036	57	0.304	15
1770	600	28	3.14	31	01093	57	0.319	16
1760	628	28	3.45	33	0.01150	59	0.335	16
1750	656	28	3.78	35	01209	59	0.351	16
1740	684	28	4.13	37	01268	61	0.367	16
1730	712	29	4.50	39	0.01329	61	0.383	17
1720	741	28	4.89	41	01390	63	0.400	16
1710	769	29	5.30	43	01453	64	0.416	17

TABLE II.—Continued.

v	$S(v)$	Diff.	$A(v)$	Diff.	$I(v)$	Diff.	$T'(v)$	Diff.
1700	798	29	5.73	45	0.01517	65	0.433	17
1690	827	29	6.18	47	01582	66	0.450	18
1680	856	30	6.65	50	01648	67	0.468	17
1670	886	29	7.15	52	0.01715	68	0.485	18
1660	915	30	7.67	54	01783	70	0.503	18
1650	945	30	8.21	56	01853	71	0.521	18
1640	975	30	8.77	58	0.01924	72	0.539	19
1630	1005	31	9.35	62	01996	74	0.558	18
1620	1036	30	9.97	64	02070	75	0.576	19
1610	1066	30	10.61	66	0.02145	77	0.595	19
1600	1096	31	11.27	69	02222	78	0.614	19
1590	1127	31	11.96	72	02300	79	0.633	20
1580	1158	31	12.68	76	0.02379	81	0.653	20
1570	1189	31	13.44	78	02460	82	0.673	20
1560	1220	32	14.22	82	02542	84	0.693	20
1550	1252	32	15.04	86	0.02626	86	0.713	21
1540	1284	32	15.90	88	02712	87	0.734	21
1530	1316	32	16.78	92	02799	89	0.755	21
1520	1348	32	17.70	95	0.02888	91	0.776	21
1510	1380	33	18.65	98	02979	93	0.797	22
1500	1413	33	19.63	100	03072	94	0.819	22
1490	1446	33	20.63	105	0.03166	96	0.841	22
1480	1479	33	21.68	109	03262	98	0.863	22
1470	1512	34	22.77	114	03360	101	0.885	23
1460	1546	34	23.91	119	0 03461	103	0.908	23
1450	1580	34	25.10	124	03564	105	0.931	24
1440	1614	34	26.34	128	03669	107	0.955	24
1430	1648	34	27.62	133	0.03776	109	0.979	24
1420	1682	35	28.95	138	03885	112	1.003	25
1410	1717	35	30.33	143	03997	114	1.028	25
1400	1752	35	31.76	149	0.04111	116	1.053	26
1390	1787	36	33.25	154	04227	119	1.079	26
1380	1823	35	34.79	160	04346	122	1.105	26

TABLE II.—CONTINUED.

v	$S(v)$	Diff.	$A(v)$	Diff.	$I(v)$	Diff.	$T(v)$	Diff.
1370	1858	36	36.39	164	0.04468	124	1.131	27
1360	1894	37	38.03	170	04592	127	1.158	27
1350	1931	36	39.73	175	04719	129	1.185	27
1340	1967	37	41.48	181	0.04848	133	1.212	27
1330	2004	37	43.29	185	04981	136	1.239	28
1320	2041	37	45.14	191	05117	139	1.267	27
1310	2078	38	47.05	196	0.05256	142	1.294	28
1300	2116	38	49.01	203	05398	144	1.322	29
1290	2154	38	51.04	212	05542	148	1.351	30
1280	2192	39	53.16	221	0.05690	152	1.381	30
1270	2231	38	55.37	230	05842	156	1.411	31
1260	2269	39	57.67	240	05998	160	1.442	31
1250	2308	40	60.07	249	0.06158	165	1.473	32
1240	2348	40	62.56	258	06323	169	1.505	33
1230	2388	40	65.14	267	06492	174	1.538	33
1220	2428	42	67.81	278	0.06666	180	1.571	34
1210	2470	42	70.59	295	06846	187	1.605	35
1200	2512	22	73.54	156	07033	97	1.640	18
1195	2534	22	75.10	160	0.07130	99	1.658	18
1190	2556	22	76.70	162	07229	100	1.676	18
1185	2578	22	78.32	165	07329	102	1.694	18
1180	2600	˙23	79.97	169	0.07431	104	1.712	19
1175	2623	23	81.66	173	07535	106	1.731	20
1170	2646	23	83.39	177	07641	108	1.751	19
1165	2669	23	85.16	182	0.07749	110	1.770	20
1160	2692	23	86.98	186	07859	113	1.790	20
1155	2715	24	88.84	190	07972	115	1.810	21
1150	2739	24	90.74	195	0.08087	117	1.831	21
1145	2763	24	92.69	199	08204	120	1.852	21
1140	2787	25	94.68	205	08324	122	1.873	22
1135	2812	25	96.73	209	0.08446	124	1.895	22
1130	2837	24	98.82	215	08570	127	1.917	23
1125	2861	25	100.97	221	08697	130	1.940	23

TABLE II.—CONTINUED.

v	$S(v)$	Diff.	$A(v)$	Diff.	$I(v)$	Diff.	$T(v)$	Diff.
1120	2886	26	103.18	226	0.08827	132	1.963	23
1115	2912	26	105.44	233	08959	135	1.986	23
1110	2938	26	107.77	239	09094	138	2.009	24
1105	2964	27	110.16	246	0.09232	141	2.033	24
1100	2991	26	112.62	251	09373	143	2.057	24
1095	3017	27	115.13	259	09516	147	2.081	25
1090	3044	27	117.72	266	0.09663	149	2.106	26
1085	3071	28	120.38	275	09812	153	2.132	26
1080	3099	28	123.13	283	09965	156	2.158	26
1075	3127	28	125.96	291	0.10121	159	2.184	26
1070	3155	29	128.87	300	10280	163	2.210	27
1065	3184	29	131.87	308	10443	166	2.237	28
1060	3213	30	134.95	317	0.10609	170	2.265	28
1055	3243	30	138.12	326	10779	173	2.293	28
1050	3273	30	141.38	338	10952	177	2.321	29
1045	3303	30	144.76	346	0.11129	181	2.350	29
1040	3333	31	148.22	355	11310	185	2.379	30
1035	3364	31	151.77	364	11495	189	2.409	31
1030	3395	32	155.41	374	0.11684	193	2.440	31
1025	3427	32	159.15	384	11877	197	2.471	31
1020	3459	32	162.99	394	12074	202	2.502	32
1015	3491	33	166.93	406	0.12276	206	2.534	32
1010	3524	33	170.99	418	12482	211	2.566	33
1005	3557	34	175.17	430	12693	215	2.599	33
1000	3591	34	179.47	443	0.12908	220	2.632	33
995	3625	35	183.90	456	13128	226	2.665	34
990	3660	35	188.46	470	13354	231	2.699	35
985	3695	36	193.16	484	0.13585	236	2.734	36
980	3731	36	198.00	498	13821	241	2.770	36
975	3767	36	202.98	513	14062	246	2.806	37
970	3803	37	208.11	529	0.14308	252	2.843	38
965	3840	37	213.40	546	14560	258	2.881	39
960	3877	38	218.86	563	14818	264	2.920	39

TABLE II.—Continued.

v	$S(v)$	Diff.	$A(v)$	Diff.	$I(v)$	Diff.	$T(v)$	Diff.
955	3915	38	224.49	580	0.15082	270	2.959	40
950	3953	39	230.29	600	15352	276	2.999	41
945	3992	39	236.29	620	15628	283	3.040	42
940	4031	39	242.49	637	0.15911	290	3.082	43
935	4070	40	248.86	657	16201	297	3.125	43
930	4110	41	255.43	676	16498	304	3.168	44
925	4151	41	262.19	698	0.16802	311	3.212	45
920	4192	42	269.17	720	17113	319	3.257	46
915	4234	43	276.37	743	17432	327	3.303	47
910	4277	43	283.80	767	0.17759	335	3.350	47
905	4320	43	291.47	793	18094	343	3.397	48
900	4363	44	299.40	819	18437	352	3.445	49
895	4407	44	307.59	845	0.18789	360	3.494	50
890	4451	45	316.04	873	19149	369	3.544	51
885	4496	46	324.77	901	19518	378	3.595	52
880	4542	47	333.78	928	0.19896	387	3.647	53
875	4589	47	343.06	961	20283	397	3.700	54
870	4636	48	352.67	997	20680	407	3.754	55
865	4684	48	362.64	1032	0.21087	418	3.809	56
860	4732	49	372.96	1064	21505	428	3.865	57
855	4781	49	383.60	1099	21933	439	3.922	58
850	4830	50	394.59	1137	0.22372	451	3.980	59
845	4880	51	405.96	1175	22823	462	4.039	61
840	4931	52	417.71	1216	23285	476	4.100	61
835	4983	53	429.87	1258	0.23761	487	4.161	63
830	5036	53	442.45	1302	24248	498	4.224	64
825	5089	54	455.47	1347	24746	511	4.288	66
820	5143	55	468.94	1395	0.25257	526	4.354	67
815	5198	55	482.89	1444	25783	540	4.421	68
810	5253	56	497.33	1495	26323	553	4.489	70
805	5309	57	512.28	1549	0.26876	568	4.559	71
800	5366	58	527.77	1604	27444	587	4.630	72
795	5424	59	543.81	1661	28031	601	4.702	74

TABLE II.—Continued.

v	$S(v)$	Diff.	$A(v)$	Diff.	$I(v)$	Diff.	$T(v)$	Diff.
790	5483	59	560.42	1722	0.28632	617	4.776	76
785	5542	60	577.64	1784	29249	634	4.852	77
780	5602	61	595.48	1849	29883	650	4.929	79
775	5663	62	613.97	1916	0.30533	670	5.008	80
770	5725	63	633.13	1988	31203	688	5.088	82
765	5788	64	653.01	2062	31891	707	5.170	84
760	5852	65	673.63	2138	0.32598	727	5.254	86
755	5917	66	695.01	2218	33325	748	5.340	87
750	5983	67	717 19	2303	34073	770	5.427	90
745	6050	68	740.22	2389	0.34843	791	5.517	91
740	6118	69	764.11	2480	35634	814	5.608	93
735	6187	69	788.91	2574	36448	837	5.701	96
730	6256	71	814.65	2673	0.37285	861	5.797	97
725	6327	72	841.38	2776	38146	887	5.894	100
720	6399	73	869.14	2882	39033	912	5.994	102
715	6472	74	897.96	2996	0.39945	940	6.096	104
710	6546	75	927.92	3115	40885	968	6.200	106
705	6621	77	959.07	3238	41853	995	6.306	109
700	6698	78	991.45	3366	0.42848	1024	6.415	111
695	6776	79	1025.2	350	43872	1054	6.526	114
690	6855	80	1060.2	364	44926	1089	6.640	116
685	6935	81	1196.6	378	0.46015	1128	6.756	119
680	7016	82	1134.4	394	47143	1159	6.875	122
675	7098	84	1173.8	409	48302	1192	6.997	125
670	7182	85	1214.7	427	0.49494	1228	7.122	127
665	7267	87	1257.4	444	50722	1267	7.249	131
660	7354	88	1301.8	463	51989	1307	7.380	134
655	7442	89	1348.1	482	0.53296	1349	7.514	137
650	7531	91	1396.3	502	54645	1392	7.651	140
645	7622	92	1446.5	523	56037	1436	7.791	143
640	7714	94	1498.8	546	0.57473	1482	7.934	147
635	7808	95	1553.4	568	58955	1529	8.081	150
630	7903	97	1610.2	592	60484	1579	8.231	154

TABLE II.—CONTINUED.

v	$S(v)$	Diff.	$A(v)$	Diff.	$I(v)$	Diff.	$T(v)$	Diff.
625	8000	98	1669.4	618	0.62063	1633	8.885	158
620	8098	100	1731.2	644	63696	1690	8.543	162
615	8198	101	1795.6	673	65386	1737	8.705	166
610	8299	103	1862.9	702	0.67123	1799	8.871	170
605	8402	105	1933.1	733	68922	1859	9.041	174
600	8507	107	2006.4	765	70781	1923	9.215	179
595	8614	108	2082.9	800	0.72704	1988	9.394	183
590	8722	111	2162.9	836	74692	2055	9.577	188
585	8833	112	2246.5	872	76747	2126	9.765	192
580	8945	114	2333.7	911	0.78873	2199	10.957	197
575	9059	116	2424.8	954	81072	2276	10.154	203
570	9175	118	2520.2	998	83348	2356	10.357	208
565	9293	120	2620.0	1043	0.85704	2440	10.565	213
560	9413	122	2724.3	1091	88144	2526	10.778	219
555	9535	124	2833.4	1142	90670	2617	10.997	225
550	9659	126	2947.6	1196	0.93287	2711	11.222	231
545	9785	129	3067.2	1252	95998	2810	11.453	237
540	9914	131	3192.4	1312	98808	2913	11.690	243
535	10045	133	3323.6	1374	1.01721	3019	11.933	250
530	10178	135	3461.0	1440	1.04740	3133	12.183	257
525	10313	138	3605.0	1509	1.07873	3247	12.440	264
520	10451	140	3755.9	1582	1.11120	3366	12.704	271
515	10591	143	3914.1	1660	1.14486	3495	12.975	279
510	10734	146	4080.1	1743	1.17981	3633	13.254	287
505	10880	148	4254.4	1829	1.21614	3779	13.541	295
500	11028	151	4437.3	1920	1.25393	3919	13.836	302
495	11179	153	4629.3	2017	1.29312	4070	14.138	312
490	11332	156	4831.0	2118	1.33382	4232	14.450	320
485	11488	160	5042.8	2226	1.37614	4399	14.770	330
480	11648	162	5265.4	2340	1.42013	4575	15.100	340
475	11810	165	5499.4	2461	1.46588	4760	15.440	350
470	11975	168	5745.5	2588	1.51348	4953	15.790	360
465	12143	172	6004.3	2724	1.56301	5157	16.150	370

TABLE II.—CONTINUED.

v	$S(v)$	Diff.	$A(v)$	Diff.	$I(v)$	Diff.	$T(v)$	Diff.
460	12315	175	6276.7	2868	1.61458	5368	16.520	382
455	12490	178	6563.5	3020	1.66826	5593	16.902	394
450	12668		6865.5		1.72419		17.296	

TABLE III.

θ	(θ)	Diff.	Tan θ	Diff.	θ	(θ)	Diff.	Tan θ	Diff.
0° 00′	0.00000	582	0.00000	582	11° 00′	0.19560	616	0.19438	604
0 20	00582	582	00582	582	11 20	20176	618	20042	606
0 40	01164	582	01164	582	11 40	20794	621	20648	608
1 00	0.01746	582	0.01746	582	12 00	0.21415	623	0.21256	608
1 20	02328	582	02328	582	12 20	22038	625	21864	611
1 40	02910	583	02910	582	12 40	22663	627	22475	612
·2 00	0.03493	583	0.03492	583	13 00	0.23290	630	0.23087	613
2 20	04076	583	04075	583	13 20	23920	633	23700	616
2 40	04659	584	04658	583	13 40	24553	636	24316	617
3 00	0.05243	584	0.05241	583	14 00	0.25189	638	0.24933	619
3 20	05827	585	05824	584	14 20	25827	641	25552	620
3 40	06412	586	06408	585	14 40	26468	644	26172	623
4 00	0.06998	587	0.06993	585	15 00	0.27112	647	0.26795	624
4 20	07585	587	07578	585	15 20	27759	650	27419	627
4 40	08172	588	08163	586	15 40	28409	654	28046	629
5 00	0.08760	589	0.08749	586	16 00	0.29063	657	0.28675	630
5 20	09349	590	09335	587	16 20	29720	660	29305	633
5 40	09939	591	09922	588	16 40	30380	663	29938	635
6 00	0.10530	592	0.10510	589	17 00	0.31043	667	0.30573	637
6 20	11122	593	11099	589	17 20	31710	671	31210	640
6 40	11715	594	11688	590	17 40	32381	674	31850	642
7 00	0.12309	596	0.12278	591	18 00	0.33055	678	0.32492	644
7 20	12905	597	12869	592	18 20	33733	682	33136	647
7 40	13502	598	13461	593	18 40	34415	686	33783	650
8 00	0.14100	600	0.14054	594	19 00	0.35101	690	0.34433	652
8 20	14700	601	14648	595	19 20	35791	695	35085	655
8 40	15301	603	15243	595	19 40	36486	699	35740	657
9 00	0.15904	605	0.15838	597	20 00	0.37185	703	0.36397	660
9 20	16509	607	16435	598	20 20	37888	708	37057	663
9 40	17116	608	17033	600	20 40	38596	713	37720	666
10 00	0.17724	610	0.17633	600	21 00	0.39309	717	0.38386	669
10 20	18334	612	18233	602	21 20	40026	722	39055	672
10 40	18946	614	18835	603	21 40	40748	728	39727	676

43

TABLE III.—CONTINUED.

θ	(θ)	Diff.	Tan θ	Diff.	θ	(θ)	Diff.	Tan θ	Diff.
22° 00'	0.41476	732	0.40403	678	33° 00'	0.69253	992	0.64941	830
22 20	42208	738	41081	682	33 20	70245	1003	65771	837
22 40	42946	744	41763	684	33 40	71248	1015	66608	843
23 00	0.43690	749	0.42447	689	34 00	0.72263	1027	0.67451	850
23 20	44439	754	43136	692	34 20	73290	1040	68301	856
23 40	45193	760	43828	695	34 40	74330	1052	69157	864
24 00	0.45953	766	0.44523	699	35 00	0.75382	1065	0.70021	870
24 20	46719	772	45222	702	35 20	76447	1078	70891	878
24 40	47491	778	45924	707	35 40	77525	1092	71769	885
25 00	0.48269	785	0.46631	710	36 00	0.78617	1106	0.72654	893
25 20	49054	791	47341	714	36 20	79723	1120	73547	900
25 40	49845	798	48055	718	36 40	80843	1134	74447	908
26 00	0.50643	805	0.48773	722	37 00	0.81977	1149	0.75355	917
26 20	51448	812	49495	727	37 20	83126	1165	76272	924
26 40	52260	818	50222	731	37 40	84291	1182	77196	933
27 00	0.53078	826	0.50953	735	38 00	0.85473	1197	0.78129	941
27 20	53904	834	51688	739	38 20	86670	1213	79070	950
27 40	54738	842	52427	744	38 40	87883	1231	80020	958
28 00	0.55580	849	0.53171	749	39 00	0.89114	1249	0.80978	968
28 20	56429	857	53920	753	39 20	90363	1266	81946	977
28 40	57286	865	54073	758	39 40	91629	1285	82923	987
29 00	0.58151	874	0.55431	763	40 00	0.92914	1303	0.83910	996
29 20	59025	882	56194	768	40 20	94217	1324	84906	1006
29 40	59907	892	56962	773	40 40	95541	1343	85912	1017
30 00	0.60799	900	0.57735	778	41 00	0.96884	1363	0.86929	1026
30 20	61699	909	58513	784	41 20	98247	1385	87955	1037
30 40	62608	919	59297	789	41 40	99632	1407	88992	1048
31 00	0.63527	928	0.60086	795	42 00	1.01039	1429	0.90040	1059
31 20	64455	939	60881	800	42 20	02468	1452	91099	1071
31 40	65394	949	61681	806	42 40	03920	1475	92170	1082
32 00	0.66343	959	0.62487	812	43 00	1.05395	1499	0.93252	1093
32 20	67302	970	63299	818	43 20	06894	1524	94345	1106
32 40	68272	981	64117	824	43 40	08418	1550	95451	1118

TABLE III.—CONTINUED.

θ	(θ)	Diff.	Tan θ	Diff.	θ	(θ)	Diff.	Tan θ	Diff.
44° 00'	1.09968	1576	0.96569	1131	52° 00'	1.57257	2522	1.27994	1547
44 20	1.11544	1604	97700	1143	52 20	1.59779	2578	1.29541	1569
44 40	1.13148	1631	98843	1157	52 40	1.62357	2638	1.31110	1594
45 00	1.14779	1660	1.00000	1170	53 00	1.64995	2701	1.32704	1619
45 20	1.16439	1690	1.01170	1185	53 20	1.67696	2764	1.34323	1645
45 40	1.18129	1720	1.02355	1198	53 40	1.70460	2831	1.35968	1670
46 00	1.19849	1751	1.03553	1213	54 00	1.73291	2900	1.37638	1698
46 20	1.21600	1784	1.04766	1228	54 20	1.76191	2971	1.39336	1725
46 40	1.23384	1817	1.05994	1243	54 40	1.79162	3045	1.41061	1754
47 00	1.25201	1852	1.07237	1259	55 00	1.82207	3122	1.42815	1783
47 20	1.27053	1887	1.08496	1274	55 20	1.85329	3201	1.44598	1813
47 40	1.28940	1923	1.09770	1291	55 40	1.88530	3285	1.46411	1845
48 00	1.30863	1960	1.11061	1308	56 00	1.91815	3371	1.48256	1877
48 20	1.32823	2000	1.12369	1325	56 20	1.95186	3460	1.50133	1910
48 40	1.34823	2040	1.13694	1343	56 40	1.98646	3553	1.52043	1943
49 00	1.36863	2081	1.15037	1361	57 00	2.02199	3650	1.53986	1980
49 20	1.38944	2124	1.16398	1379	57 20	2.05849	3751	1.55966	2015
49 40	1.41068	2168	1.17777	1398	57 40	2.09600	3856	1.57981	2052
50 00	1.43236	2214	1.19175	1418	58 00	2.13456	3965	1.60033	2092
50 20	1.45450	2260	1.20593	1438	58 20	2.17421	4079	1.62125	2131
50 40	1.47710	2309	1.22031	1459	58 40	2.21500	4197	1.64256	2172
51 00	1.50019	2360	1.23490	1479	59 00	2.25697	4321	1.66428	2215
51 20	1.52379	2412	1.24969	1502	59 20	2.30018	4450	1.68643	2258
51 40	1.54791	2466	1.26471	1523	59 40	2.34468	4585	1.70901	2304
					60 00	2.39053	4726	1.73205	2351

45

www.ingramcontent.com/pod-product-compliance
Lightning Source LLC
Chambersburg PA
CBHW021806190326
41518CB00007B/465